The Electronic Structure and Chemistry of Solids

The Electronic Structure and Chemistry of Solids

P. A. COX

Fellow of New College and
Lecturer in Inorganic Chemistry,
University of Oxford

OXFORD NEW YORK TOKYO
OXFORD UNIVERSITY PRESS

Oxford University Press, Walton Street, Oxford OX2 6DP
Oxford New York
Athens Auckland Bangkok Bombay
Calcutta Cape Town Dar es Salaam Delhi
Florence Hong Kong Istanbul Karachi
Kuala Lumpur Madras Madrid Melbourne
Mexico City Nairobi Paris Singapore
Taipei Tokyo Toronto
and associated companies in
Berlin Ibadan

Oxford is a trade mark of Oxford University Press

Published in the United States
by Oxford University Press Inc., New York

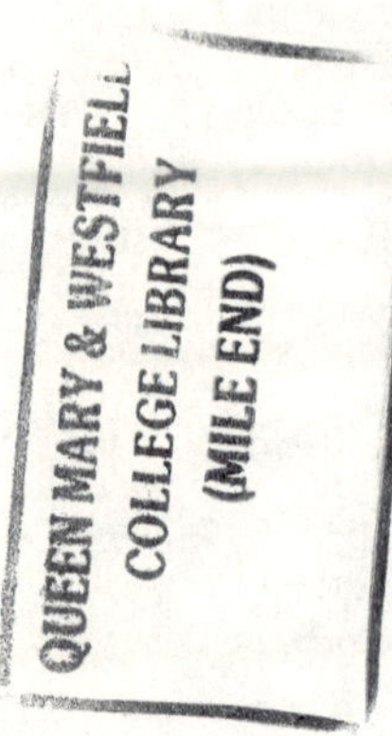

First published 1987
Reprinted 1989, 1991, 1993, 1995

British Library Cataloguing in Publication Data
Cox, P. A.
The electronic structure and chemistry of solids.
1. Solid state chemistry
I. Title
541'.0421 QD478

Library of Congress Cataloging in Publication Data
Cox, P. A.
The electronic structure and chemistry of solids.
Bibliography: p.
Includes index.
1. Solid state chemistry. I. Title.
QD478.C69 1987 541'.042 86–23462

ISBN 0 19 855204 1 (Pbk)

Printed and bound in Great Britain by
Biddles Ltd, Guildford and King's Lynn

Preface

For many years the study of the electronic structure and properties of solids has been regarded as the preserve of physicists. Chemists, who spend a good deal of time and effort looking at electronic structures of molecules, have tended to avoid these aspects of solids. The situation is changing however, and there has recently been an upsurge of interest in those aspects of solid-state chemistry that have to do with electronic properties and their relation to chemical bonding. In part, this change is due to the discovery of new classes of solids with unusual and surprising properties. Many of these investigations have been carried out by teams of chemists and physicists working together, and such collaboration is slowly helping to break down the traditional barriers between these two disciplines. In spite of these developments, there are still considerable problems of communication, which arise largely because chemists are not educated in solid-state concepts. The many excellent solid-state physics books are rather unrewarding for the average chemist, since in the first place they have too mathematical an approach, and secondly they tend to stop short of discussing the more complex solids that chemists find interesting. This book is an attempt to bridge the gap, and to show that the electronic structures of solids can be understood using ideas that are more familiar to chemists. The emphasis is on a fairly descriptive approach with relatively few mathematical derivations, but with plenty of chemically interesting examples. Modern spectroscopic techniques, especially photoelectron spectroscopy, are given a prominent role, as they provide very direct information, of a chemically relevant kind, about electronic energy levels in solids.

The first three chapters form a fairly elementary, self-contained account of the electronic structure of simple solids, and could be used in a first course on this subject, suitable for undergraduate students with a reasonable basic grounding in inorganic chemistry. As I have indicated in the text, one or two small sections, dealing with dielectric properties, could be omitted from an elementary course. The later chapters contain slightly more specialised material, intended for graduate or advanced undergraduate students. The formal treatment of band theory is left until this point, since I feel that this topic is not essential for a basic understanding of the chemical aspects of electronic structure. The band model is introduced through the LCAO or tight-binding method: not only is this a much more obvious approach for chemists who already have some acquaintance with simple molecular orbital theory, but also it is more useful for the majority of chemical applications. Although the concepts of band theory are rather unfamiliar to chemists, it should be possible to read this chapter without too much theoretical background. An elementary understanding of complex numbers is required, and some familiarity with valence theory would be useful. There is no doubt, however, that many students will find this the most difficult chapter. Although some of the concepts developed there are referred to in later chapters, the detailed theory is not necessary for these. Chapter 4 could therefore be omitted on a first reading of the book. In fact, band theory is of limited use for many chemically interesting solids, and the last three chapters discuss situations where the model breaks down. These include many topics of current research interest to both chemists and physicists, such as metal–insulator transitions, mixed-valency compounds, one-dimensional conductors and 'molecular metals', and surface and defect properties.

This book has grown out of lecture courses, both at undergraduate and graduate levels, given in Oxford in the last few years. I am grateful to all those—students and colleagues—who have attended the courses, and whose comments, both critical and encouraging, have helped to clarify my ideas about how best to present the material. Other colleagues have contributed indirectly through many stimulating discussions in their specialist fields. The influence of both Peter Day and John Goodenough is particularly marked in the choice of topics in the latter part of the text. I must also thank those who read all or part of the manuscript: these include Allen Hill, who provided a valuable non-specialist viewpoint, and the reviewers chosen by Oxford University Press, who made detailed and extremely useful comments on the first draft. Finally, I would like to express by appreciation, both to the staff of Oxford University Press, and to my wife Christine and the children, for encouraging and supporting me in what they have felt to be a worthwhile achievement.

Oxford P. A. C.
1986

Contents

A note on units

In conformity with modern practice, all formulae involving electrostatics and magnetism have been written using SI units. Readers more familiar with c.g.s. electrostatic units may simply make the substitution $\varepsilon_0 = 1/4\pi$. All energy values in the text—and in most diagrams—have been quoted in electron volts (eV). The only exception to this is in a number of electronic absorption spectra taken from the literature, where the more common spectroscopist's unit, the wave number (cm^{-1}), has been retained. Values for these units, and for fundamental constants referred to in the text, are listed below:

e	electronic charge	1.602×10^{-19} C
m	free electron mass	9.110×10^{-31} kg
ε_0	dielectric permittivity of vacuum	8.854×10^{-12} Fm^{-1}
μ_0	magnetic permeability of vacuum	$4\pi \times 10^{-7}$ Hm^{-1}
μ_B	Bohr magneton	9.274×10^{-24} Am^2
eV	electron volt	1.602×10^{-19} J
		or 8065 cm^{-1}
k	Boltzmann constant	1.318×10^{-23} J K^{-1}
		8.617×10^{-5} eV K^{-1}
h	Planck's constant	6.626×10^{-34} J S
$\hbar$	$h/2\pi$	1.055×10^{-34} J S

1
Introduction

1.1 The importance of solids

Most of the chemical elements and their compounds are solids at room temperature. The study of solids, and of the factors which determine their structures and properties, therefore forms an important part of chemistry. It is the electronic theory of valency which provides the basis for our modern understanding of chemistry, and this represents one important reason for studying the electronic structure of solids. Another motivation for this study is that many of the characteristic properties of solids depend directly on the behaviour of the electrons in them. Among these properties are the following:

Electrical conduction: the properties of metals and semiconductors.
Optical properties: the emission and absorption of light, and photochemical reactions.
Magnetic properties: paramagnetism; ferro- and antiferro-magnetic ordering.
Surface properties: especially those involved in the transfer of electrons to another phase, such as in electrochemistry.

Many of the applications of solids depend on these electronic properties, which form the basis for integrated circuits, lasers, magnetic recording tape, and solar energy converters, to mention just a few. Although we shall look briefly at some applications, the main aim of this book is to discuss the basic principles involved in studying and understanding the electronic structure. We shall see that many features of the electronic structure of a solid can be understood using the same ideas of chemical bonding which are familiar in molecules. Indeed, the wide diversity of electronic properties of solids reflects an equally wide range of chemical-bonding interactions. One of the major aims of the chapters which follow is to explore this relationship, and to show how chemical ideas can be used to obtain important insights into the properties of solids. Some aspects of electronic structure are certainly unique to solids, and these will involve concepts which are rather unfamiliar to chemists. As far as possible, however, we shall try to treat a solid as a 'very large molecule', and to explain the special features of solids by an extension of the electronic theory of molecules.

Most solids form **crystals**, in which the atoms or molecules are packed in regular arrays. In much of the following, it will be assumed that we are dealing with crystalline solids. There is however an important class of solids which are to some extent **amorphous**, and lack the long-range order of a crystal. The traditional methods for treating the electronic structure of a solid depend heavily on crystalline order, and break down when faced with highly disordered materials. The advantage of the chemical approach is that it emphasizes the importance of the **local bonding** of an atom in determining electronic structure. In many amorphous solids, such as the familiar silicate glasses, the atoms retain to a large extent the same local environment as in crystals. Thus many of the important features of electronic structure are not strongly affected by the disorder. For this reason it is only in Chapter 4, which discusses the band theory of solids, that we shall have to take explicit account of long-range crystalline order. Crystalline solids, however, contain **defects** and impurities, which are associated with a change in the local environment of atoms. The same is true at the **surface** of a solid, where the coordination of atoms is lowered from the bulk values. Defects, impurities and surfaces, may therefore have a strong influence on the electronic properties of a solid. Because of the importance of these effects in the applications of solids, we shall consider them in detail in Chapter 7.

1.2 Chemical classification of solids

From a chemical point of view, solids are most conveniently classified in terms of the types of force which hold the atoms together. The four important classes of simple solid are:

molecular
ionic
covalent
metallic

and some examples of each are shown in Table 1.1. The interactions, and the general types of structure that they give rise to, are discussed in the sections which follow.

1.2.1 Molecular solids

These are solids composed of individual atoms or molecules which retain their identity, being held by relatively weak van der Waals' forces. They are thus characterized by low boiling points and sublimation energies. For non-polar atoms and molecules, the principal interaction is the London dispersion force, caused by the interaction of fluctuating dipoles that arise from the motion of

Table 1.1
Classes of simple solid

Class	*Example*
Molecular solids	Xe
	N_2
	benzene (C_6H_6)
	$HgCl_2$
Ionic solids	NaCl
	MgO
	CaF_2
Covalent solids	C (diamond)
	P
	SiO_2
	GaAs
Metallic solids	Na
	Fe
	Cu

the electrons in each molecule. An approximate expression for the energy of attraction between two atoms or molecules at a distance R is:

$$E_{\text{disp}} = -(3/16\pi\varepsilon_0)h\omega_0\alpha^2/R^6 \tag{1.1}$$

where α is the electronic polarizability, and $h\omega_0$ an average electronic excitation energy. Since the dispersion force is non-directional and falls off rapidly with distance, the main structural driving force is the packing of molecules as closely together as possible. Thus fairly symmetrical groups such as atomic xenon and molecular nitrogen have close-packed structures like those of metals. With molecules such as benzene that are far from spherical, the close-packing in the structure is not so obvious. A view down the crystal axis containing the molecular plane of benzene (see Fig. 1.1) shows molecules at two different orientations. Calculations have shown that this is indeed the arrangement giving the closest packing of molecules, whilst avoiding the repulsive non-bonding interactions between the hydrogen atoms.

Polar molecules have more directional forces: for example the electrostatic interaction between permanent dipoles, which gives an attraction between the positive end of one dipole, and the negative end of a neighbouring one. The most extreme examples of such directional forces occur with **hydrogen bonding**. Figure 1.1(b) shows how the structure of boric acid, $B(OH)_3$, contains sheets of molecules hydrogen-bonded together. Hydrogen bonding can also

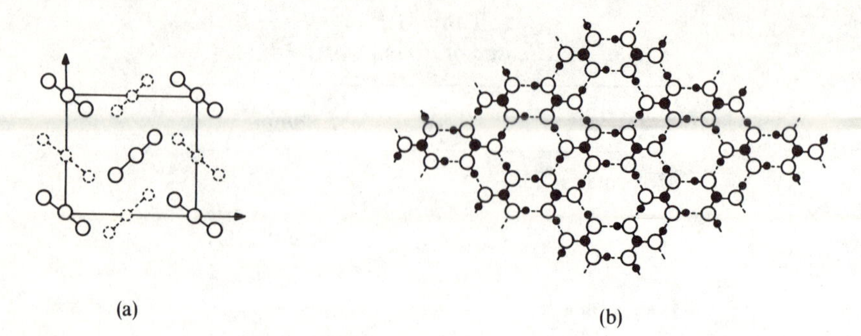

Fig. 1.1 Molecular lattices. (a) Benzene: view in the molecular plane (from Adams 1974). (b) Hydrogen-bonded structure of boric acid, $B(OH)_3$ (from Wells 1984.)

occur in conjunction with the other interactions discussed below, particularly in ionic lattices containing water or hydroxide groups.

1.2.2 Ionic solids

In the ionic model, the bonding is assumed to arise from the electrostatic attraction between ions (which are formed from the transfer of electrons from one atom to another). The energy of interaction between two ions, with charges z_1 and z_2, is given by the Coulomb law:

$$E_{\text{Coul}} = z_1 z_2 e^2/(4\pi\varepsilon_0 R). \tag{1.2}$$

The $1/R$ dependence makes this a long-range force, so that it is not sufficient to consider the interactions between near-neighbours alone. For example, in the rock-salt structure, illustrated in Fig. 1.2, each cation has six nearest neighbour anions at a distance r, twelve cations at $\sqrt{2}r$, eight anions at distance $\sqrt{3}r$, and so on. The binding of an ion depends on its Coulomb interaction with all these successive shells of ions, and hence on the infinite series:

$$-e^2(6-12/\sqrt{2}+8/\sqrt{3}-\ldots)/(4\pi\varepsilon_0 r). \tag{1.3}$$

The term in brackets is known as the **Madelung constant,** A_M. The series does not converge as it is written, and cannot be estimated by simple summation. Madelung constants for structures such as rock-salt, which have a high degree of symmetry, can be calculated by grouping the ions into neutral blocks, and obtaining a series which converges quickly. The value for the rock-salt structure is 1.747 56. . . . For complex crystal structures, however, more sophisticated mathematical methods are required.

The lattice energy in the ionic model can be calculated by balancing the attractive Coulomb force against the repulsive interaction that results from the short-range overlap of the closed-shell ions. An approximate formula often

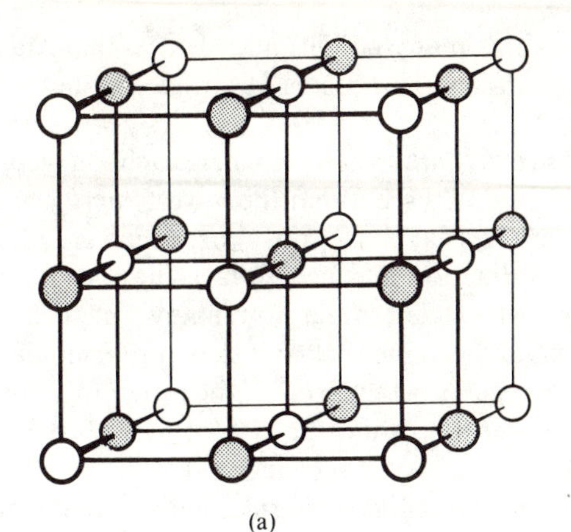

(a)

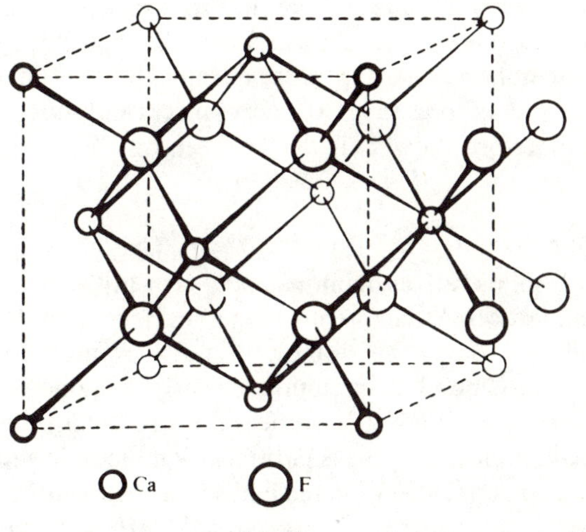

(b)

Fig. 1.2 Ionic lattices. (a) Rock-salt (NaCl). (b) Fluorite (CaF_2) (from Wells 1984.)

used for repulsive forces is:

$$E_{rep} = B/R^n \tag{1.4}$$

where B and n are constants for a given pair of ions. n is normally in the range 6–9. The final form of the lattice energy is found by adding the attractive and repulsive terms, and differentiating to give the minimum energy, which must correspond to the observed distance. This allows the constant B to be eliminated, and gives:

$$U_{lat} = -Nz_1z_2e^2A_M(1-1/n)/(4\pi\varepsilon_0 r). \tag{1.5}$$

For ionic solids such as many halides and oxides, this equation is found to agree well with experimental estimates of lattice energies, obtained from the Born–Haber cycle.

The Coulomb force favours structures where each ion is surrounded by ones of opposite charge. It would seem that the lowest energy structures would be the ones with the largest Madelung constants, which are generally those with highest possible coordination number. In fact, the coordination is limited by geometrical factors which determine how many ions can be packed round another one. For example, in the rock-salt structure each ion is surrounded by six others forming a regular octahedron. If the ratio of the ion sizes is less than about 0.414, the smaller ion cannot be in contact with the larger ones, and a smaller interionic distance could be achieved in a structure with lower coordination number. The **radius ratio rules** which result from this consideration are a useful qualitative guide to ionic structures, but they are not by any means obeyed perfectly, even in the simplest solids such as the alkali halides. There are many reasons for such a breakdown, the most important probably being that ions do not act as hard spheres of fixed radius: the interionic distance depends on a balance of long-range attractive forces and short-range repulsive ones, and will therefore show a change between different structures.

1.2.3 Covalent solids

The structures of solids such as diamond and quartz (SiO_2) are often described as giant covalent lattices. Atoms are held together by covalent bonds which are essentially similar to those in small molecules. For example, the bond energy and length of the C–C bond in diamond are nearly the same as those found in alkanes. The structures of covalent lattices are also determined by the same factors as those in molecules. This is particularly noticeable with the elements of the non-metal Groups IV to VI in the Periodic Table. The first-row elements carbon, nitrogen, and oxygen are exceptional: carbon forms the graphite structure as well as diamond, and the other two have molecular lattices with diatomic N_2 and O_2 molecules. This exceptional behaviour is due to the tendency for multiple bonding between first-row elements, as displayed in many of their simple compounds. In the lower rows, single bonding is the rule. Silicon, germanium and tin in Group IV have the tetrahedral diamond structure. The Group V elements have structures with pyramidal three-coordination, and those of Group VI are two-coordinate. Some examples of these structures are shown in Fig. 1.3. The so-called **8-N rule** relates the coordination in the solid to the group number N, and is a reflection of the structures found in many small covalent molecules, such as hydrides. The progressive lowering of coordination number occurs as the number of electrons increases beyond that of the available atomic orbitals. In Group V elements for example, it is more favourable for one electron pair to be

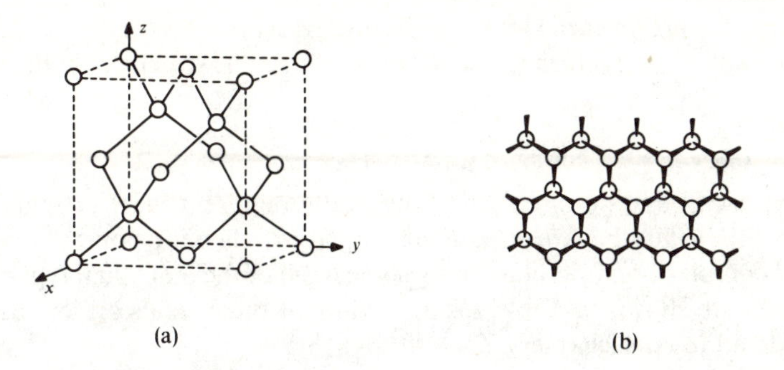

Fig. 1.3 Covalent solids. (a) Diamond. (b) Layer structure of Group V elements P, As, Sb (from Wells 1984.)

accommodated in a non-bonding orbital, so that only three electrons are available to form bonds with neighbouring atoms.

1.2.4 Metallic solids

Metals are characterized by a delocalized sharing of electrons between many atoms. The same kind of delocalization is found on a small scale in cluster compounds formed by many metallic elements. In metallic solids, however, the delocalization of valence electrons extends throughout the solid.

In general, metallic structures are formed by elements with a deficiency of electrons, relative to the number of bonding orbitals. As in electron-deficient molecules, the maximum stability can be achieved, not by using electrons in localized bonds between pairs of atoms, but by sharing out the electrons as widely as possible. Metals tend to adopt close-packed or nearly close-packed structures, some examples of which are shown in Fig. 1.4. The detailed factors which determine the structure of a metal, however, are quite complicated, and depend on the slightly different energy distributions of bonding orbitals

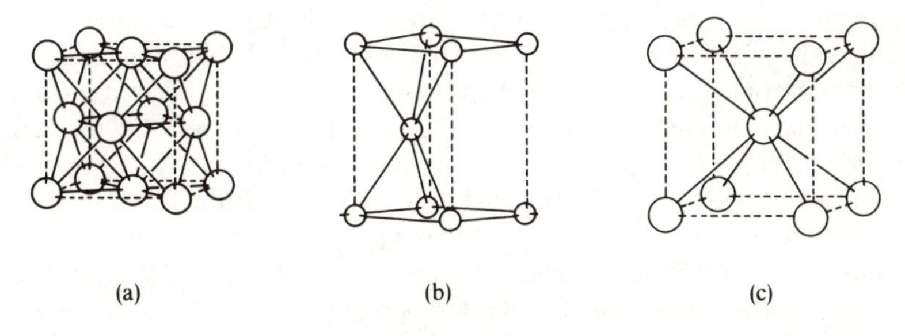

Fig. 1.4 Metallic structures. (a) Cubic close packed (face-centred cubic). (b) Hexagonal close packed. (c) Body-centred cubic.

which are formed in each structure. To discuss these requires the use of band theory, and we shall return to the problem of metallic structures at the end of Chapter 4.

1.2.5 More complex solids

The types of bonding discussed above really represent ideal extreme cases. Many of the solids that are most interesting from a chemical point of view have a more complex bonding, combining some or all of the individual types. Some examples are shown in Table 1.2, and many of these solids are discussed in more detail in later chapters.

Table 1.2
Some solids which combine different types of bonding

Bonding type	*Example*
Intermediate ionic–covalent	CdS TiO_2 CsAu
Intermediate ionic–covalent plus van der Waals	CdI_2
Ionic, metallic	NbO TiO
Ionic, metallic, van der Waals	ZrCl
Ionic, covalent, metallic	$K_2Pt(CN)_4Br_{0.3}3H_2O$[a]
Covalent, metallic, van der Waals	graphite
Ionic, covalent, metallic, van der Walls.	TTF: TCNQ[b]

[a] Structure shown in Fig. 6.1 on p. 167.
[b] Shown in Fig. 6.9 on p. 176. TTF = tetrathiafulvavene; TCNQ = tetracyanoquinodimethane.

In many compounds, such as cadmium sulphide, CdS, the difference of electronegativity between the metal and non-metal is such that there must be a fair degree of ionic character in the bond, although the charge transfer can be by no means complete. Thus we would describe such a solid as being intermediate between ionic and covalent. The same is true in compounds containing ions with a high formal charge, as in TiO_2. Although we shall find the ionic model to be useful in discussing the electronic structure of compounds such as this, it is certain that the ionic description does not give a true picture of the charge distribution. A surprising example in this category is caesium auride, CsAu, which although a compound of two metallic elements, is itself non-metallic, and is sometimes formulated Cs^+Au^-.

Sometimes intermediate ionic–covalent bonding is combined with other interactions. For example, in the layer structure of compounds such as cadmium iodide, CdI_2, shown in Fig. 1.5, there are adjacent planes of iodine atoms, held together by van der Waals' forces.

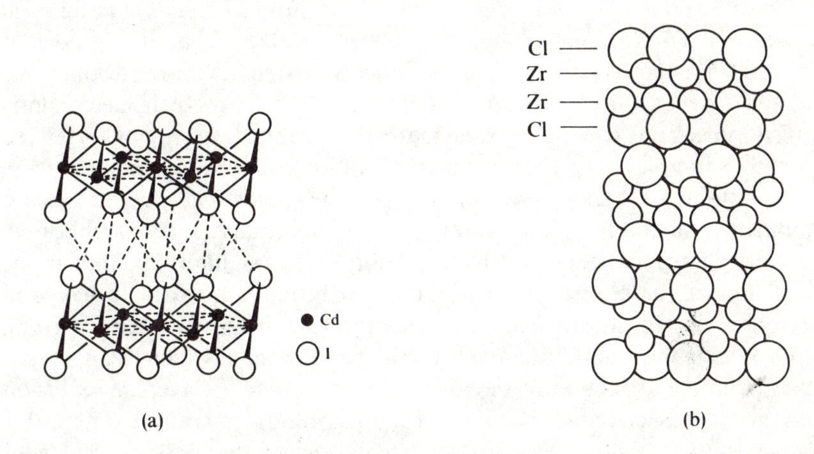

Fig. 1.5 Layer compounds. (a) Cadmium iodide structure (from Wells 1984). (b) Double layer structure of ZrCl (from R. E. McCarley, in D. B. Brown (ed.), *Mixed-valence compounds,* D. Reidel, 1980.)

Interesting electronic properties often arise in solid compounds where some degree of metallic bonding is combined with other types. This happens, for example, in transition-metal compounds, especially where the metal has a low oxidation state, and has electrons left over which can form metal–metal bonds. Such compounds may form three-dimensional structures such as the oxides TiO and NbO, but sometimes the metallic bonding may dominate the structure, giving chains or sheets of metal atoms. In the structure of ZrCl, shown in Fig. 1.5(b), the double layers of zirconium atoms have metallic bonding between them. Chlorine is presumably held by an ionic bond as Cl^-, and the different layers must also be held together by van der Waals' forces. Delocalized metallic bonding may even occur in predominantly molecular solids, such as the compound TTF: TCNQ, formed between tetrathiafulvalene (TTF) and tetracyanoquinodimethane (TCNQ). The electronic properties of this solid, discussed in Chapter 6, show that there is some charge transfer between the two components, giving it also a degree of ionic bonding.

These examples show that the bonding in interesting solids may be quite complex, and often difficult to treat satisfactorily by 'first principles' theoretical methods. We shall see, however, that it is often possible to use 'chemical intuition' to isolate the interactions that are important from the point of view of electronic properties.

1.3 Electrons in solids

1.3.1 Orbitals in atoms, molecules, and solids

The chemist's picture of bonding in molecules and solids is based heavily on the concept of orbitals. The idea originates with the atomic orbitals coming from the solution of Schrödinger's equation for the hydrogen atom. It is important to realise that the use of the orbital model in systems with more than one electron depends on certain approximations. Because of their electrostatic repulsion, electrons tend to avoid each other, so that their motion is correlated in a highly complex way. Fairly accurate solutions of Schrödinger's equation can be obtained for many-electron atoms, using numerical methods with a computer. The mathematical form of these solutions is so complicated, however, that they do not generally lead to any useful qualitative picture. In the orbital model, the electron repulsion is treated in an approximate way, and electrons are assumed to move independently. Thus in a many-electron atom, each orbital is the solution of a Schrödinger equation, in which the potential comes from the attraction to the nucleus, balanced by the average repulsion from the other electrons in their own orbitals. Although it is approximate, this idea leads to a highly successful interpretation of the Periodic Table of Elements. Since the basic principles used to interpret atomic structure are also applied to molecules and solids, it is useful to summarize them here:

(i) The **Pauli exclusion principle** requires that only two electrons, with opposed spins, may occupy each orbital.

(ii) It is necessary to know the **degeneracy** of each orbital, and its **relative energy**. In atoms there is one *s* orbital at a given energy, three *p* orbitals at the same energy, five *d* orbitals, and so on. The different screening of orbitals in many-electron atoms leads to the energy order:

$$1s < 2s < 2p < 3s < 3p < 4s < 3d < \ldots$$

(iii) These considerations lead to the **Aufbau** method for constructing the ground-state electron configuration of an atom. (*Aufbau* is a German word, meaning 'building up'.) Electrons are placed, two at a time, in orbitals of increasing energy. The structure of the Periodic Table of Elements follows as a natural consequence of the energies and degeneracies of the different atomic orbitals.

(iv) When a degenerate set of orbitals is left partially filled, the arrangement of electrons is determined by **Hund's rules**. The most important rule is the first, which specifies that electrons should be placed, so far as possible, in different orbitals with their spins parallel. This is the arrangement that minimizes the electrostatic repulsion between electrons.

The same ideas are used in molecular orbital (MO) theory. Although very accurate mathematical representations of molecular orbitals may be obtained with large computers, it is usual to approximate the orbitals in molecules as **linear combinations of atomic orbitals:** the so-called LCAO method. The simplest case is that of hydrogen, H_2, where we form molecular orbitals as linear combinations of the 1*s* atomic orbitals. The electron density, given as the square of the wave function, must by symmetry be the same on each atom. The possible combinations of the atomic orbitals χ_A and χ_B on the two atoms are:

$$\psi_1 = \chi_A + \chi_B \tag{1.6}$$

and

$$\psi_2 = \chi_A - \chi_B. \tag{1.7}$$

The electron density in these orbitals is pictured in Fig. 1.6, and it can be seen that ψ_1 has an increase of density in the region between the nuclei, where there is a favourable attractive potential. On the other hand, in ψ_2 the electrons tend to avoid this favourable region. As a consequence, ψ_1 is a **bonding** molecular orbital, which leads to an energy lowering if it is occupied by electrons. ψ_2 is a corresponding **antibonding** orbital, and has an energy higher than that of the isolated atoms. More detailed calculations have shown that the bonding in H_2 is slightly more complicated than suggested by this simple picture. The atomic orbitals actually contract somewhat on forming the bonding MO, leading to an increase of electron density close to the nucleus. The case of hydrogen, which has no inner-shell electrons, is probably exceptional however, and in most cases covalent bonding can be described in terms of a charge build-up, due to overlap of atomic orbitals, in the internuclear region.

Molecular orbitals may be constructed in a similar way from the atomic orbitals of heavier atoms. The bonding interaction depends on the degree of overlap between the atomic orbitals on adjacent atoms. To combine effectively to give molecular orbitals, two atomic orbitals must have:

(i) a fairly similar energy;

(ii) the correct relative symmetry. Thus in linear molecules there is no interaction between σ type orbitals, which have their maximum density along the molecular axis, and π type orbitals, which have a nodal plane along the axis, and lobes of opposite sign either side.

The relative energy of atomic orbitals is important in heteronuclear molecules, formed from atoms of more than one kind. Figure 1.6 shows a simple case, where atoms A and B have one atomic orbital each. The molecular orbitals may now be written:

$$\psi_1 = a_1\chi_A + b_1\chi_B \tag{1.8}$$

and

$$\psi_2 = a_2\chi_A - b_2\chi_B. \tag{1.9}$$

The bonding orbital ψ_1 is concentrated more on atom B, which has the atomic orbital of lower energy. When this orbital is occupied by two electrons, there will be a net transfer of charge from A to B. We shall study this situation applied to solids in more detail in Chapter 3. The degree of mixing between the A and B atomic orbitals depends inversely on their energy separation. In the limit, when the atomic levels are very different, the molecular orbitals correspond closely to the original atomic orbitals. The ground-state charge distribution is now ionic: A^+B^-. We call B the more **electronegative** atom, because it has the ability to attract electrons to it in the molecule. A rough guide to electronegativity may be obtained from the atomic ionization energy. In Fig. 1.6, the ionization energy is clearly greater for atom B, since its atomic orbital energy is lower (that is, more negative) than that of A.

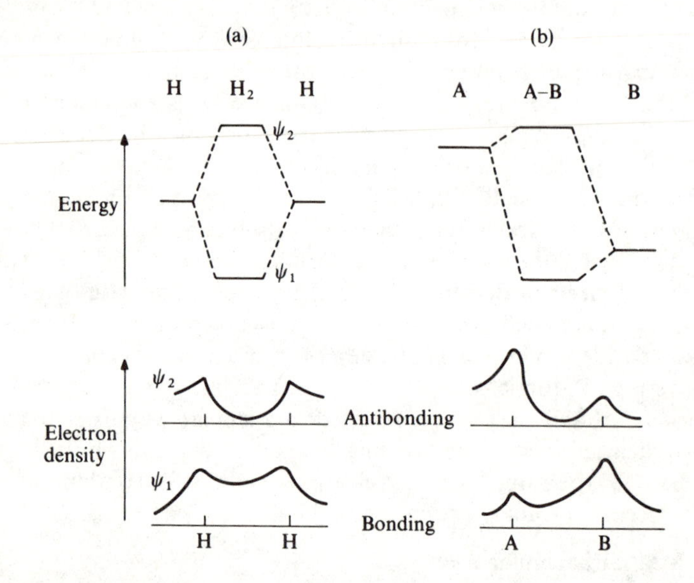

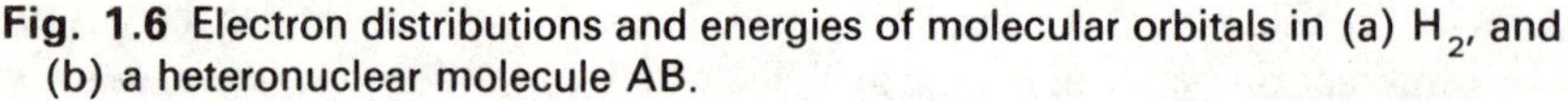
Fig. 1.6 Electron distributions and energies of molecular orbitals in (a) H_2, and (b) a heteronuclear molecule AB.

It is normal to distinguish between **core** and **valence** atomic orbitals. The core orbitals are composed of the previously filled shells. Generally they are of smaller radius than the valence orbitals, and their overlap in molecules is small. Thus core orbitals have very little bonding influence. Sometimes the distinction between core and valence orbitals is less clear. In the lanthanide elements, for example, the partially filled 4*f* shell must be considered as a valence orbital in terms of its energy. On the other hand, it is highly contracted in size, and has virtually no overlap with surrounding atoms. From this point of view, the 4*f* orbitals should be considered as part of the core.

In polyatomic molecules, a greater variety of molecular orbitals can be formed. The MO theory emphasizes the delocalized nature of the electron distribution, so that molecular orbitals are generally extended over all the constituent atoms. The total number of molecular orbitals formed—which may be bonding, antibonding, or non-bonding—is the same as the number of valence atomic orbitals used to make them. As the molecules become larger, their molecular orbitals therefore become more numerous, and more closely spaced in energy (see Fig. 1.7). We can think of a solid as no more than a very large molecule. The orbitals which are extended throughout the solid can be called **crystal orbitals**, and their properties will be described in detail in Chapter 4. Strictly speaking, in a finite solid, there can only be a finite number of crystal orbitals, equal to the number of valence atomic orbitals. Since there are now so many orbitals however, we can for most purposes neglect the energy spacing between them, and assume that they form continuous **bands** of energy levels. This is shown in Fig. 1.7.

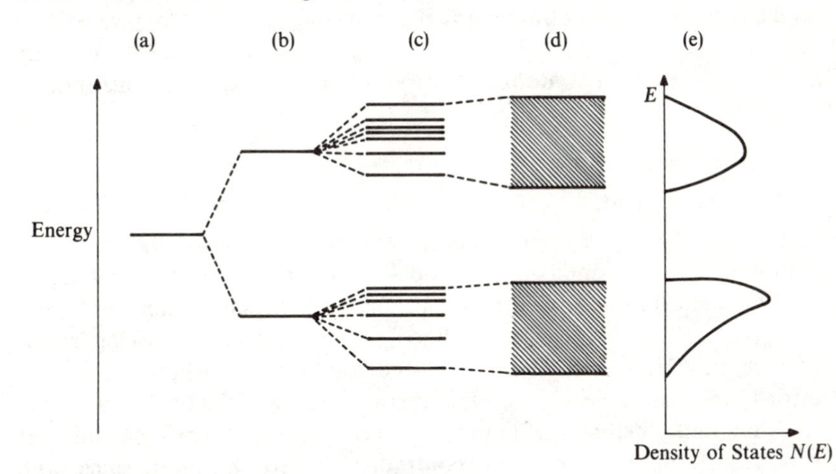

Fig. 1.7 Orbital energies of (a) atom, (b) small molecule, (c) large molecule, (d) solid, and (e) density of states corresponding to (d).

As our imaginary molecule becomes larger and larger, the molecular orbitals will not be distributed uniformly over all energies. Figure 1.7 shows some energy regions where there are no orbitals, and which therefore correspond to a gap between the bands of the solid. Even within the allowed bands, more orbitals are concentrated together at some energies than at others. This leads to the concept of the **density of states** $N(E)$, which is defined by:

> $N(E)\,\mathrm{d}E$ is the number of allowed energy levels per unit volume of the solid, in the energy range E to $E+\mathrm{d}E$.

$N(E)$ is zero in the forbidden band gaps.

In many parts of this book, the energy levels of a solid will be represented by a simple block diagram showing the band energies, as in Fig. 1.7(d). There are times, however, when the more detailed information contained in the density of states curve is useful. Diagrams such as this can be regarded as the natural generalization of molecular orbital diagrams to the case of a solid. For elementary purposes we do not need to know the detailed form of the crystal orbitals (for example the coefficients of the different atomic orbitals), and this problem will not be discussed until Chapter 4.

The electronic properties of a solid depend on the energies of the bands, their widths, and the gaps between them, and in Chapter 3 we shall see how these can be related to the chemical bonding in different types of solid. It is useful to realize at this point, however, that the width of a band depends, like the splitting between bonding and antibonding molecular orbitals, on the degree of interaction between neighbouring atoms. Valence *s* and *p* orbitals, especially in the metallic elements on the left-hand side of the Periodic Table, overlap strongly to give bands which are many electron volts in width. Contracted core orbitals, on the other hand, give very narrow bands (less than 0.1 eV wide). These orbitals retain their atomic identity in solids, and do not contribute to the bonding.

1.3.2 Bands and bonds

At first sight there is a contradiction between the view of the electronic structure in terms of bands of orbitals delocalized throughout the solid, and the more chemical picture of electrons localized on particular atoms or bonds. This apparent paradox is not confined to solids. When thinking about molecules, we often use the idea of localized bonding or non-bonding pairs of electrons, and yet molecular orbital theory gives a different picture of a polyatomic molecule, in which electrons occupy orbitals which extend over several atoms. There is really no contradiction here, however, since both descriptions are valid ways to picture the very complicated many-electron wave function. It is convenient to think of the distribution more simply by splitting it into contributions from individual electrons or pairs of electrons. Since electrons are indistinguishable however, the way in which we split up the total distribution is quite arbitrary.

It may be helpful to think of the simple case of two helium atoms in contact. One view of the electron distribution would be to consider two electrons in each of the 1*s* atomic orbitals, that is, to write the electron configuration as:

$$(1s_A)^2(1s_B)^2.$$

Alternatively, we could use molecular orbitals, formed as the linear combinations given in equations 1.6 and 1.7. The electron configuration is then given by two electrons each in the bonding and antibonding orbitals:

$$(\psi_1)^2(\psi_2)^2.$$

The wave functions corresponding to these two configurations are in fact **identical**. In the first one, we have split the distribution into electron pairs which are localized on atoms, whereas in the second we have used orbitals which are delocalized between the atoms. In more complex cases, we can make any independent set of linear combinations of filled orbitals, without in any way changing the total wave function. For some purposes, especially when considering the bonding in the ground state, the localized description is more useful. However, when we deal with excited states, it is generally better to use the delocalized orbitals, which give the energy bands in a solid. In many cases, the important electronic properties of a solid are best interpreted using the band picture, and in subsequent chapters we shall explore in more detail the relationship between this model and the more 'chemical' view of localized bonding.

The example discussed above shows that when we think in terms of energy bands, the orbitals which make up these bands are precisely the ones which we would use in the localized picture. In an ionic solid, the localized view would be of atomic orbitals on the anions and cations. The bands in an ionic solid are therefore made of combinations of these same atomic orbitals. For example, there will be occupied bands formed from the filled anion orbitals, and empty bands from the empty cation orbitals. In a covalent solid, we can similarly think of the bands as combinations of orbitals which are either bonding or antibonding between near-neighbour atoms. These ideas will be developed further in Chapter 3. The idea of bands in molecular solids is generally less useful, since the interactions within a molecule are much stronger than those between molecules. There are some molecular solids however, where the electronic properties *are* dominated by inter-molecular interactions, and then the best picture is to think of the bands as made up of combinations of the orbitals on different molecules. Thus the occupied bands will be made up of combinations of filled molecular orbitals, and the empty bands from combinations of empty orbitals.

The equivalence of localized and delocalized pictures only holds for non-metallic solids with completely filled bands. There is no way in which the electron distribution in a metal can be represented in terms of electrons localized on individual atoms, or in individual bonds. The same is true in many molecular cases, for example in metal-cluster compounds or boron hydrides, where we are forced to use bonding orbitals which extend over many atoms. Another example is the π bonding in benzene, for which no simple localized picture is possible. We can think of benzene using the rather artificial concept of **resonance** between the two localized Kekulé structures, but the delocalized molecular orbital picture is generally better. Some early attempts were made to apply the resonance idea to metals, by considering the large number of possible localized bonds which could be formed between atoms. However, this model becomes very complicated, and does not lead to any simple interpretation of the properties of a metal. The band picture, with crystal orbitals delocalized through the solid, is the best one to use. Electrons in partly filled orbitals do not always delocalize in this way, however. When the overlap of atomic orbitals is small, and the bands narrow, it may be energetically more favourable for electrons to remain localized on their individual atoms. This happens with many compounds of transition metals and lanthanides, which in spite of having partially filled *d* or *f* orbitals, are not metallic. The band model breaks down in this situation, which because of its importance in a large number of chemically interesting solids, is discussed in detail in Chapter 5.

1.4 Metals, insulators, and semiconductors

1.4.1 Metallic and non-metallic solids

The most obvious electronic property of a solid is that of **electrical conductivity**. Metallic solids conduct down to the lowest attainable temperatures. In non-metallic solids on the other hand, there may be some conductivity at higher temperatures, but it declines as the temperature is lowered. The different behaviour can be understood in terms of the filling of the bands, and is illustrated in Fig. 1.8. We shall see in Chapter 4 that for each crystal orbital corresponding to the motion of the electron in one direction, there is another at the same energy, corresponding to the reverse motion. In a filled band, therefore, the net motion of the electrons cancels, and there can be no conductivity. A solid with a filled band, and an energy gap to the next, empty band, will therefore be an insulator in its ground state. A metal however has a partially filled band, with no energy gap above the top-occupied level. Under the influence of an electric field, electrons near the top-filled level can move into

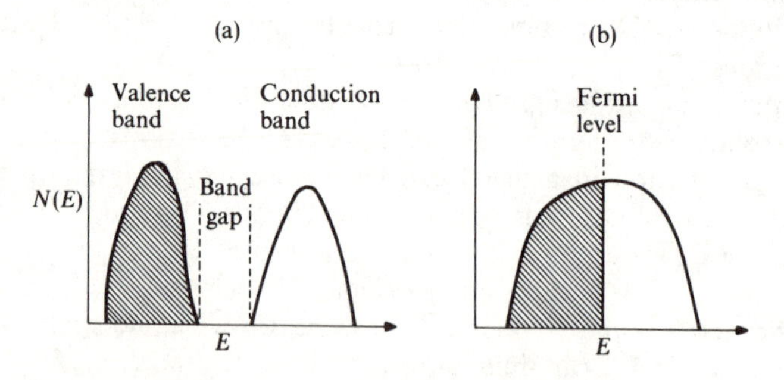

Fig. 1.8 Densities of states in (a) non-metallic solid, and (b) metal. Shading represents occupied levels.

orbitals where they give rise to a net motion of charge in the solid, so that an electric current flows.

Confirmation of this picture comes from spectroscopic measurements of different kinds, which are described in more detail in Chapter 2. Apart from infra-red absorption due to vibrations of the atoms, a non-metallic solid generally does not absorb radiation below a certain threshold energy. In an atom or molecule, the absorption of light is associated with the excitation of an electron from an occupied to an empty orbital (see Fig. 1.9). Although molecular absorption bands may appear broad when measured in solution, this is because of the vibrations which may be excited simultaneously with the electronic transition. The MO levels themselves are sharp, as can be seen from gas-phase spectra. In a solid, the threshold is followed by a broad region of absorption, showing that the filled and empty electronic levels are themselves broad. Simple absorption spectroscopy such as this is a way of measuring the **band gap**: that is, the energy gap between the top-filled band, known as the **valence band**, and the bottom empty band, the **conduction band**. Measured band gaps range from more than 12 eV in some ionic solids, down to 0.1 eV or less in some semiconductors (1 eV corresponds to a photon with a wave-number of about 8065 cm^{-1}; the visible spectrum extends from photon energies of 1.5 eV to 3 eV). The optical properties of a metal are quite different, since there is no energy gap, and the range of electronic excitation extends

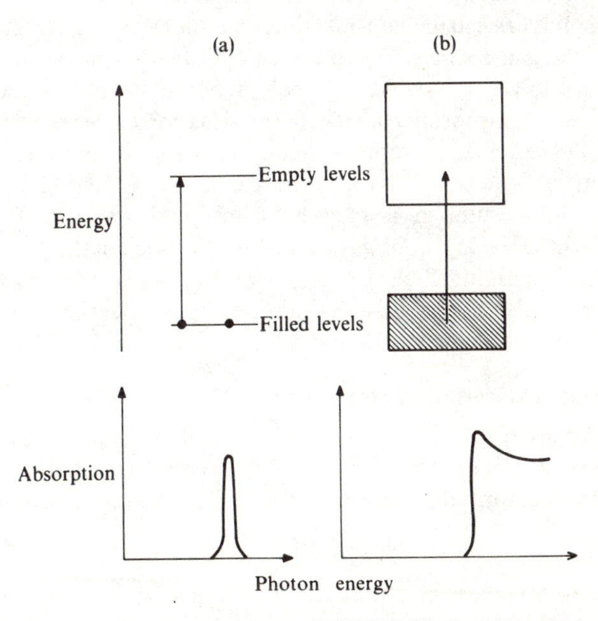

Fig. 1.9 Spectroscopic transition between the top filled and bottom empty levels in (a) an atom or molecule, and (b) a non-metallic solid.

down to zero energy. In fact, metals show high **reflectivity** of light. This is a consequence of the strong interaction of radiation with the relatively free electrons, and is described more quantitatively in Section 2.4 of Chapter 2.

Solids with large band gaps are highly insulating under normal conditions. In an ionic solid such as NaCl or MgO, the conductivity measured at high temperatures arises mostly not from mobile electrons, but from **ions** migrating through the lattice. Some solids such as AgI can have ionic conductivities comparable to those of electrolyte solutions, but the values are still low compared with the conductivity of a metal. Electronic conductivity in a non-metallic solid requires some electrons to be excited to the conduction band. Excitation can be produced optically, by absorption of a photon of energy greater than the band gap. This is the phenomenon of **photoconductivity**, which is used, for example, in photocopiers (see Section 7.4.1). If the band gap is small, however, electrons may be thermally excited into the conduction band, as discussed in the following sections.

Using the relationship between the band picture and chemical bonding models, discussed in the previous section, we can understand in simple terms why many solids are non-metallic. In the ionic model, the valence band is made up of the occupied anion atomic orbitals, and the conduction band from the cation orbitals. In ionic solids such as NaCl or MgO, the anion levels are completely filled (Cl^- has the $(3p)^6$ configuration, for example), and the cation orbitals empty. Thus we have a gap between a filled band and an empty one. Molecular solids are usually non-metallic, and this can similarly be understood in terms of the gap between the top filled MO (which makes up the valence band) and the lowest empty MO (which gives rise to the conduction band). Interesting electronic properties occur in situations where orbitals remain partially occupied. For example, in many transition-metal compounds, the transition-metal ion is left with some electrons remaining in d orbitals. Sometimes this results in a metallic compound, as with ReO_3, where rhenium(VI) has the $(5d)^1$ configuration. As mentioned in the previous section however, the remaining d electrons in transition-metal compounds do not always form bands of delocalized orbitals, as the simple theory suggests they should.

1.4.2 Thermal excitation of electrons

Many properties of solids depend on the thermal excitation of electrons from the ground state. The number of atoms or molecules in excited states at a given temperature T is normally described by the Boltzmann distribution:

$$n_i \propto \exp(-E_i/kT) \tag{1.10}$$

where E_i is the energy of a state, and k is the Boltzmann constant. The assumptions made in deriving the Boltzmann distribution are not applicable to electrons in a solid. In fact, it is necessary to take account of two properties of electrons:

(i) They obey the exclusion principle, so that each state (when the spin direction is specified as well as the orbital) can only hold one electron.

(ii) Electrons are totally indistinguishable, so that an exchange of electrons between occupied levels does not lead to a different arrangement.

As is shown in Appendix A1, these properties give rise to the **Fermi–Dirac** distribution:

$$f(E) = \frac{1}{1 + \exp[(E - E_F)/kT]} \tag{1.11}$$

The function $f(E)$ gives the fraction of the allowed levels with energy E which are occupied, and is shown plotted for different temperatures in Fig. 1.10. At absolute zero, the Fermi–Dirac distribution corresponds, as we would expect, to a sharp cut-off between completely filled levels below the energy E_F, and completely empty levels above it. This energy is called the **Fermi energy** or **Fermi level**, and in a metal is simply the top-filled level in the band. At higher temperatures, the distribution is smeared out, showing that some electrons are excited thermally to higher energies.

There are many properties of metals, such as their specific heat, that depend indirectly on the Fermi–Dirac distribution, but it is also possible to measure it directly, using the technique of photoelectron spectroscopy. As is described in

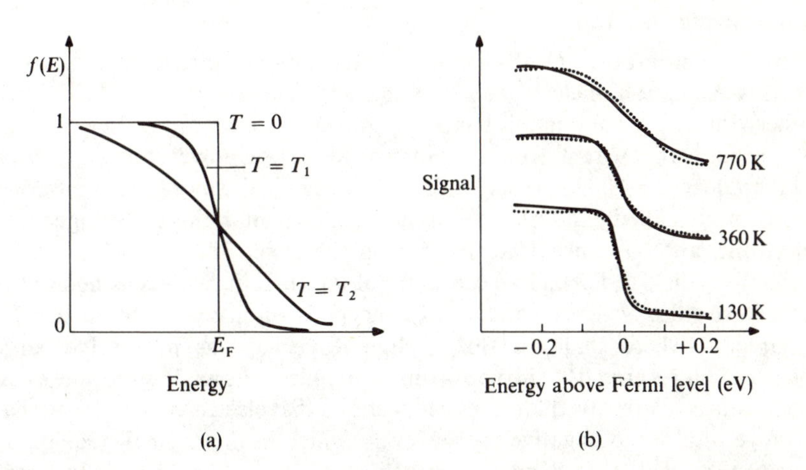

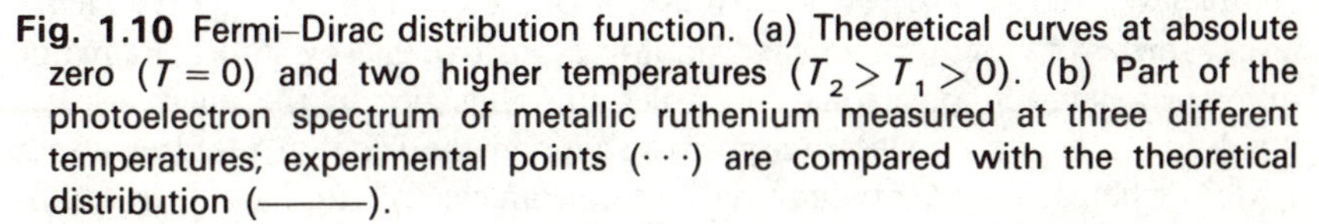

Fig. 1.10 Fermi–Dirac distribution function. (a) Theoretical curves at absolute zero ($T = 0$) and two higher temperatures ($T_2 > T_1 > 0$). (b) Part of the photoelectron spectrum of metallic ruthenium measured at three different temperatures; experimental points (· · ·) are compared with the theoretical distribution (———).

Chapter 2, this type of spectroscopy gives a direct measurement of the energy distribution of occupied levels in a molecule or solid. Some spectra of a crystal of metallic ruthenium, taken at different temperatures, are shown in Fig. 1.10(b). The increasing width of the distribution as the temperature is raised agrees closely with the prediction of equation 1.11. The total width of the distribution, from an energy where nearly all levels are filled, to one where they are nearly empty, is around $4kT$. At room temperature this has the value of about 0.1 eV. Since the total width of the bands is generally several electron volts, we can see that the number of electrons which are thermally excited is quite a small fraction of the total. This fact is important in understanding many properties of metals: for example, the contribution made by conduction electrons to the specific heat is much smaller than the classical equipartition law would predict.

Although at temperatures above absolute zero the boundary between filled and unfilled levels is no longer sharp, the Fermi energy still has a fundamental importance, as the thermodynamic **chemical potential** for electrons in the solid. The normal conditions for chemical equilibrium between two phases show that the chemical potentials of any mobile species must be the same. This is true for the Fermi levels of solids placed in electrical contact. Electrons will flow from one solid to the other, causing a potential difference across the contact. When the solids are in electrical equilibrium, the contact potential must be just sufficient to make the Fermi levels of both solids the same.

1.4.3 Semiconductors

Semiconductors are non-metallic solids that conduct electricity by virtue of the thermal excitation of electrons across an energy gap. Electrons excited into an otherwise empty conduction band may move under an applied electric field and hence carry current. Conductivity may also arise, however, from electrons in the valence band, when this is not completely filled. Since the net motion of electrons in a filled band is zero, it is most convenient in this case to ignore the electrons, and to concentrate instead on the small number of unoccupied orbitals in the band. Thus we can think of the current carriers as **holes** in the otherwise filled valence band. This concept is discussed in more detail in Chapter 4, where shall see that such holes behave as positively charged particles. Holes are rather like positrons, and indeed the analogy is quite a close one. Dirac's relativistic quantum theory shows that electrons in free space have a range of allowed negative energy levels, which in the normal 'vacuum' are completely filled. A positron is essentially a unfilled hole in the negative energy levels, formed by exciting an electron into a positive energy state. The major difference between positrons and holes in solids lies in the energy scales involved: to create an electron–positron pair in the vacuum requires about 1 MeV, whereas the energy required to create an electron and a hole in a non-metallic solid is equal to the excitation energy across the band gap. In a typical semiconductor, this may be 1 eV or less.

In the last section, we saw that the thermal excitation of electrons is governed by the Fermi–Dirac distribution function, equation (1.11). To apply this equation to a non-metallic solid, we must first locate the Fermi level E_F. At low temperatures, E_F represents the boundary between filled and empty levels, and in a non-metallic solid, it must therefore be somewhere in the gap between the valence and conduction bands. This is illustrated in Fig. 1.11, and is quite different from the case of a metal, where the Fermi level lies within the partially occupied band. Figure 1.11(a) corresponds to a perfectly stoichiometric solid where there are no electrons or holes in the ground state. At any temperature, the number of electrons excited into the conduction band must be the same as the number of holes left in the valence band. Since $f(E)$ in equation 1.11 gives the fraction of occupied levels, the actual number of electrons at a particular energy can be found by multiplying $f(E)$ by the number of allowed levels, that is the density of state function $N(E)$. In Fig. 1.11, the densities of states in the valence and conduction bands are assumed to be equal, and in order for the number of electrons in the conduction band to be equal to that of the holes in

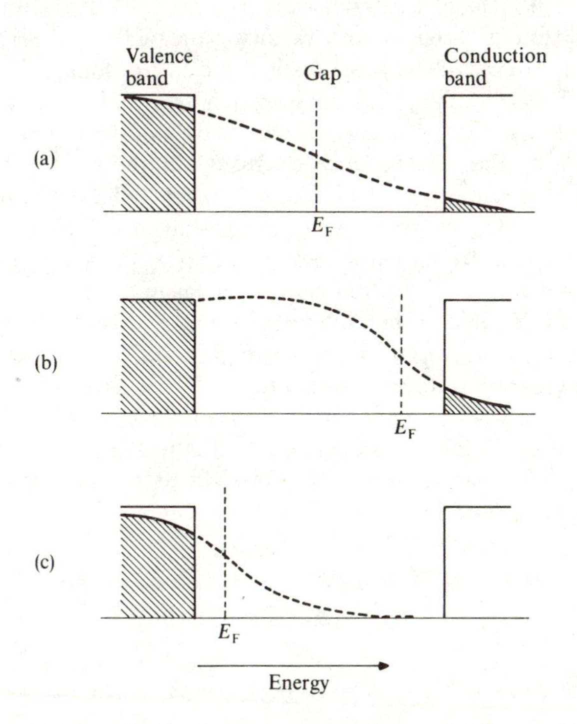

Fig. 1.11 Fermi–Dirac distribution in a semiconductor. (a) Pure solid; (b) *n*-type; and (c) *p*-type semiconductors. Shading represents occupied levels.

the valence band, the Fermi level must be placed mid-way in the energy gap. When the densities of states in the two bands are not equal, it is necessary to shift E_F slightly, but this is normally a small correction. In a pure solid, the Fermi level is usually close to the middle of the band gap.

For an energy E at the bottom of the conduction band, we have:

$$E - E_F = E_g/2$$

where E_g is the band gap. Unless the gap is very small, $E - E_F$ is usually much larger than kT. Under these conditions the exponential term dominates the bottom of equation 1.11, and the number of excited electrons is given by:

$$n \propto \exp(-E_g/2kT). \tag{1.12}$$

The conductivity should therefore show Arrhenius-like behaviour, with an activation energy equal to half the band gap. This is indeed found to be the case, with very pure semiconductors. However, the *intrinsic* properties of the pure solid may be difficult to measure. For example, elemental silicon has a band gap of 1.1 eV, and the concentration of electrons and holes calculated at room temperature is around 10^9 per cm^3. As we shall see in Chapter 7, impurities or defects in a crystal can give rise to extra electrons in the conduction band or holes in the valence band. Clearly, only a very low concentration of extra charge carriers is sufficient to dominate the electrical properties. The effect of impurities is shown in Figs. 1.11(b) and (c). An **n-type** semiconductor has more electrons in the conduction band than holes in the valence band, so that the *negative* electrons are the predominant current carriers. A **p-type** semiconductor has more *positive* holes in the valence band.

It can be seen in the diagram how the extra electrons or holes cause a shift of the Fermi level, away from its ideal mid-gap position. In an n-type semiconductor, E_F is closer to the conduction band, whereas in the p-type case, it moves down towards the valence band. Semiconductors in solid-state devices are deliberately doped with impurities. The shift in Fermi level caused by the extra carriers is just as important for the operation of these devices as the additional conductivity. The properties and applications of semiconductors are discussed in some detail in Chapter 7. Semiconduction is found in many compounds, especially those of transition metals, where the extra electrons or holes can result from small deviations from perfect stoichiometry.

Further reading

The general chemical background to this book is covered in a number of textbooks, notably:

J. E. Huheey (1983). *Inorganic chemistry* (3rd edn). Harper and Row.

C. S. G. Phillips and R. J. P. Williams (1965). *Inorganic chemistry* (two volumes). Oxford University Press.

F. A. Cotton and G. Wilkinson (1980). *Inorganic chemistry* (4th edn). John Wiley and Sons.

Of the above, the first two contain chapters specifically on solids. Much more detail on structural aspects of solids is to be found in the following:

A. F. Wells (1984). *Structural inorganic chemistry* (5th edn). Oxford University Press.

D. M. Adams (1974). *Inorganic solids.* John Wiley and Sons.

A. R. West (1984). *Solid state chemistry and its applications.* John Wiley and Sons.

The book by Wells—in successive editions—is virtually the definitive work on structural solid state chemistry.

R. McWeeny (1979). *Coulson's valence* (3rd edn). Oxford University Press.

J. M. Murrell, S. F. A. Kettle, and J. M. Tedder (1985). *The chemical bond* (2nd edn). John Wiley and Sons.

These two good accounts of chemical valence theory both contain short chapters on solids. Books devoted to electronic aspects of solids are generally written from a physicist's point of view. An elementary, well-balanced, account is:

B. R. Coles and A. D. Caplin (1976). *Electronic structure of solids.* Edward Arnold.

A more substantial, though somewhat idiosyncratic account, is presented by:

W. A. Harrison (1980). *Electronic structure and the properties of solids.* Freeman.

Finally, excellent accounts of all aspects of solid state physics, including electronic structure and properties, are to be found in several books, including:

C. Kittel (1976). *Introduction to solid state physics* (5th edn). John Wiley and Sons.

N. W. Ashcroft and N. D. Mermin (1976). *Solid state physics.* Holt, Rinehart, and Winston.

J. S. Blakemore (1985). *Solid state physics* (2nd edn). Cambridge University Press.

R. J. Elliott and A. F. Gibson (1982). *Introduction to solid state physics and its applications.* Oxford University Press.

H. M. Rosenberg (1972). *The solid state* (2nd edn). Oxord University Press.

2
Spectroscopic methods

2.1 Introduction

The most direct source of information on the electronic energy levels of atoms, molecules, and solids, is provided by spectroscopic techniques of different kinds. Such information has played an important part in the development of the quantum theory of matter. In the chapters which follow, we shall make quite extensive use of spectroscopic results. It is useful at this point, therefore, to discuss some of the techniques, and the type of information that they can provide.

The most widely known type of electronic spectroscopy is that of optical absorption in the visible or ultraviolet (UV) range. The absorption peaks result from the transition of an electron from filled to empty orbitals—or bands in the case of solids. We have already seen in Chapter 1 that electronic absorption spectroscopy can be used to determine the energy gap between the valence and conduction bands in a non-metallic solid. However, the details of the absorption spectrum depend on both filled and empty bands, and it is relatively difficult to disentangle individual features of the separate levels. There are other methods, using photons in the X-ray region, or combining measurements of photon and electron energies, that give more direct information about individual bands. These techniques will therefore be discussed before that of optical absorption.

Table 2.1 shows a range of spectroscopic methods, including some involving measurement of electron energies as well as photons. A schematic diagram of some of the techniques is given in Fig. 2.1. The zero of energy in this figure is the vacuum level—that is, the energy at which an electron is just able to escape from the solid. The energy scale includes levels in which the electron is bound in the solid (binding energy being the energy below the vacuum level), and levels of positive energy in which the electron can enter or leave the solid. The most powerful source of information on filled levels is that of photoelectron spectroscopy, where the excess kinetic energy (the energy above the vacuum level) of electrons leaving the solid is measured. The inverse photoelectron experiment (Fig. 2.1(c)) gives similar information about empty levels, although this technique has only recently been developed, and so far has been applied to

Table 2.1
Spectroscopic methods

Method	Particles used: In	Particles used: Out	Information obtained about:
Photoelectron	Photon	Electron	Filled levels
Inverse photoelectron	Electron	Photon	Empty levels
X-ray emission	–	Photon	Filled levels
X-ray absorption	Photon	–	Empty levels
Visible/UV absorption	Photon	–	Band gap; defects
Electron energy loss	Electron	Electron	Conduction electrons

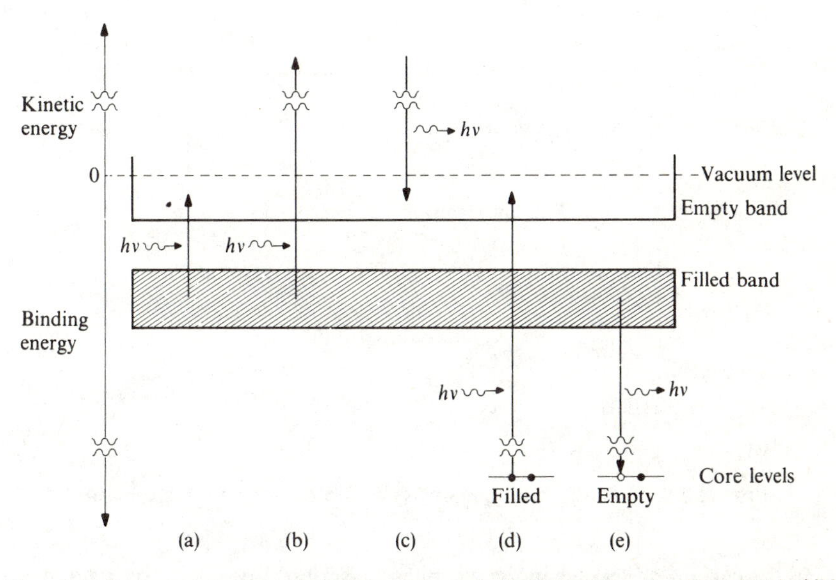

Fig. 2.1 Spectroscopic techniques. (a) optical absorption in the visible/UV range; (b) photoelectron spectroscopy; (c) inverse photoelectron spectroscopy; (d) X-ray absorption; (e) X-ray emission. Electrons with energies above the vacuum level can enter or leave the solid; in techniques (b) and (c) the scale shows the kinetic energy measured in a vacuum outside the solid.

fewer systems. These electron spectroscopy methods are discussed first in this chapter, followed by the methods in which X-ray photons are used to observe transitions to or from inner-shell orbitals (Fig. 2.1(d) and (e)). There is another technique, which measures the loss in kinetic energy of electrons scattered from a solid. This can give information similar to, but in some ways complementary to, optical absorption. It is described briefly in the final section, in connection with the optical properties of metals.

2.2 Photoelectron spectroscopy (PES)

2.2.1 General principles

The essence of the photoelectron experiment is to expose the sample—which can be a solid, or atoms or molecules in the gas phase—to a beam of monoenergetic photons that have sufficient energy to ionize electrons from the sample (see Fig. 2.2). The kinetic energies of these electrons are measured in the spectrometer, and as can be seen from Fig. 2.1(b), the kinetic energy (KE) is related to the binding energy (BE) of the electron in the sample by the formula:

$$\mathrm{KE} = h\nu - \mathrm{BE} \tag{2.1}$$

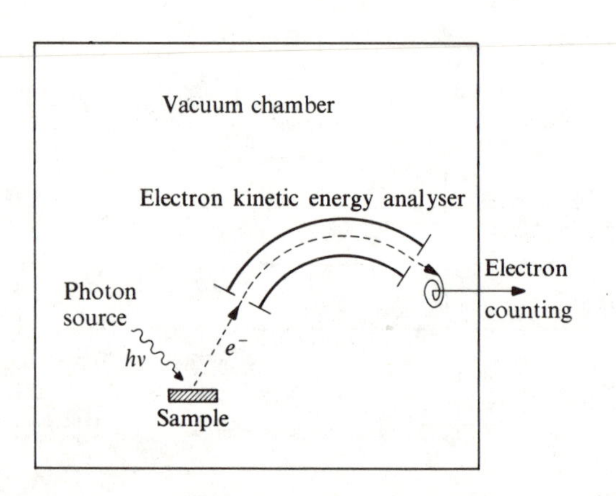

Fig. 2.2 Schematic arrangement of a photoelectron spectrometer.

In the case of molecules in the gas phase, a series of peaks is observed in the photoelectron (PE) spectrum, each corresponding to electrons ionized from a particular molecular orbital. The peaks may however have some fine structure or broadening due to vibrations excited during the ionization process. In a solid, where the electronic energy levels form more-or-less broad bands, the widths of the photoelectron peaks reflect this bandwidth. Thus in simple cases, PES is able to provide a direct measure of the absolute binding energies and the widths of various filled bands in a solid. In principle the PE spectrum can give a direct picture of the filled density of states, although it is also necessary to take account of the fact that different orbitals may have different probabilities of ionization, or so-called **ionization cross-sections**. The intensities of bands in the PE spectrum will therefore be weighted according to the cross-sections of the atomic orbitals which contribute to them. Atomic orbital ionization cross-sections vary with photon energy in ways which are now well known, and important information about the atomic orbital composition of electronic levels can sometimes be obtained by measuring spectra at different photon energies. This will be illustrated in the next section.

Traditionally, the photons used in PES are generated by monochromatic line sources. For example atomic inner-shell transitions are excited in X-ray tubes, and a magnesium anode, producing photons at 1253 eV, is often used. Electric discharges through gases give photons in the UV energy range, also from atomic emission lines. The most commonly used UV source is a helium discharge lamp, which has its strongest line at 21.2 eV. Radiation produced by highly accelerated electrons in a synchrotron can also be used for PES, although since this source gives a continuum of photon energies, the radiation must first be monochromated. This is done by diffraction, using a grating for UV photons, or a crystal for shorter wavelength photons in the X-ray region.

One important aspect of PES must be mentioned before we look at some specific examples. The electrons produced in ionization, with kinetic energies in the range 10–1000 eV depending on the photon energy, are strongly scattered by the solid. This scattering gives a strong background to the spectra, from electrons that have lost energy before escaping from the solid sample. More seriously, it means that the electrons contributing to the spectrum come mostly from a very thin surface layer, in the region of 1 or 2 nm. The PES technique is therefore used widely to study the surfaces of solids. When information about the **bulk** electronic structure of the solid is required, it is essential for the sample to be properly clean, with a surface composition and structure the same as the bulk. Cleanliness at an atomic level is often difficult to achieve, and to keep a surface free from contamination requires the use of ultra-high vacuum conditions in the spectrometer, with pressures in the range of 10^{-10} millibar. Fortunately the PE technique itself provides a means of measuring the surface composition, since photons in the X-ray range excite electrons from core levels, with binding energies characteristic of a given element. Thus X-ray photoelectron spectroscopy (XPS) is an important method for the surface analysis of solids.

2.2.2 Applications of PES

It is interesting first to compare the spectrum of a solid with that of the same species in the gas phase. Figure 2.3 shows the PE spectrum of gaseous benzene and that of its molecular solid. The bands in the gas-phase spectrum come from the filled molecular orbitals of benzene. For example, the band at around 9 eV binding energy is from the top-filled π MO. The structure on each band is due to vibrations excited in the ionization process: according to the Franck–Condon principle, the simultaneous vibrational excitation is a reflec-

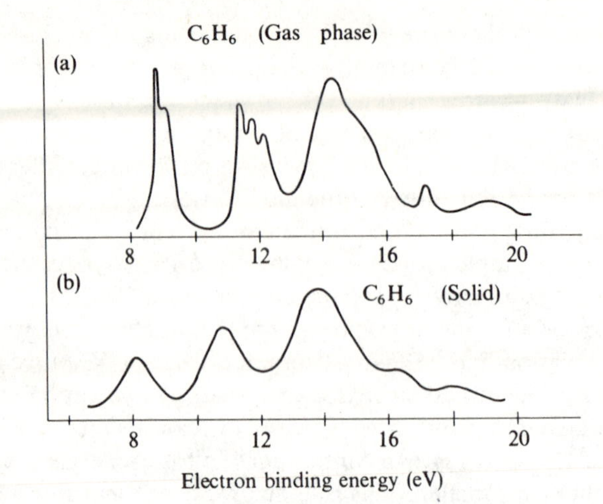

Fig. 2.3 UV photoelectron spectrum of benzene (a) in gas phase, (b) solid sample. Note the shift in energy between the two spectra. (After E. E. Koch and W. D. Grobmann: in *Photoemission in Solids*, vol. II, L. Ley and M. Cardona (eds.), Springer-Verlag, 1979, p. 269.)

tion of the changes in molecular geometry that occur when electrons are removed from the different orbitals. The solid state spectrum is very similar to the gas-phase one. As would be expected, the filled levels in the molecular solid correspond closely to the MOs of the individual molecules. However, the spectra show two noticeable differences, which must result from putting the molecules in a solid. Firstly the PES bands are broadened somewhat in the solid state spectrum. One contribution to such broadening is probably from vibrational modes of the crystal lattice that are excited in ionization. Most lattice modes are of very low energy, so that they cannot be resolved individually in the PE spectrum. As with electronic spectra measured in solution, they give rise to a general broadening. There must also be some effect, however, from the overlap of orbitals in the solid, which gives rise to bands of electronic levels, rather than sharp molecular orbitals.

The second interesting difference between the two spectra in Fig. 2.3 is that there is a shift in the absolute binding energies, those of the solid being 1.15 eV lower than in the gas phase. This is an effect which, as we shall see later, is very important in many electronic processes in solids. Even in isolated atoms and molecules, a detailed calculation of electron binding energies must take account of the response of other electrons to the ionization process. The orbitals change shape somewhat, to draw the remaining electrons closer to the vacant hole. Such an intra-molecular effect contributes to gas-phase and solid state spectra equally. But in the solid, the positive hole left on ionization will also cause the polarization of electrons in surrounding molecules, which will lower the energy required to produce the hole. An approximate estimate of the change in binding energy can be made by considering the solid as a continuum with a relative dielectric constant ε_r resulting from the molecular polarizability. If a charge q, spread out over a sphere of radius r, is moved from the vacuum into the solid, the electrostatic polarization gives an energy change equal to:

$$\Delta E_{pol} = -q^2/(8\pi\varepsilon_0 r)(1 - 1/\varepsilon_r) \tag{2.2}$$

For charges localized over atomic or molecular dimensions, the polarization energy is of the order of 1 eV.

Turning from the case of a molecular solid to a metal, Fig. 2.4 shows the PE spectrum of aluminium. Superimposed on a rising background of scattered electrons is a band of about 12 eV width, coming from the conduction band of the metal. Since only occupied levels are observed in PES, the spectra of metals show a cut-off at the Fermi energy, the top-filled level. We have already seen in Fig. 1.10 (p. 19) that PE spectra measured under high energy resolution can show how electrons are thermally excited according to the Fermi–Dirac distribution function. (Note that Fig. 1.10 was plotted with the energy scale running in the opposite direction to that in the PE spectra shown here. This was done to show more clearly the comparison with the Fermi–Dirac distribution function, also plotted.) We shall see in Chapter 3 that the density of states in a simple metal is predicted to have a functional form closely similar to that revealed by the spectrum in Fig. 2.4.

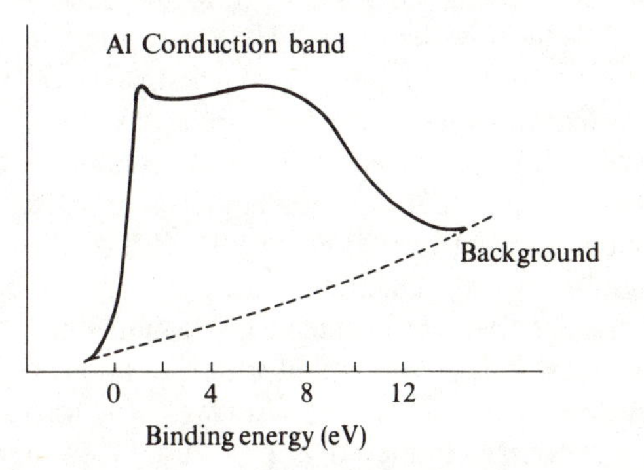

Fig. 2.4 Photoelectron spectrum of aluminium metal. (After P. Steiner, H. Hoechst, and S. Huefner: in *Photoemission in Solids*, vol. II, L. Ley and M. Cardona (eds.), Springer-Verlag, 1979, p. 369.)

The final example here shows the different information that can be obtained from PE spectra measured in the UV and X-ray energy ranges (UPS and XPS respectively). Figure 2.5 shows XPS and UPS of a more complex and interesting solid, the sodium tungsten bronze, $Na_{0.7}WO_3$. X-radiation has sufficient photon energy to ionize electrons from inner-shell or core orbitals,

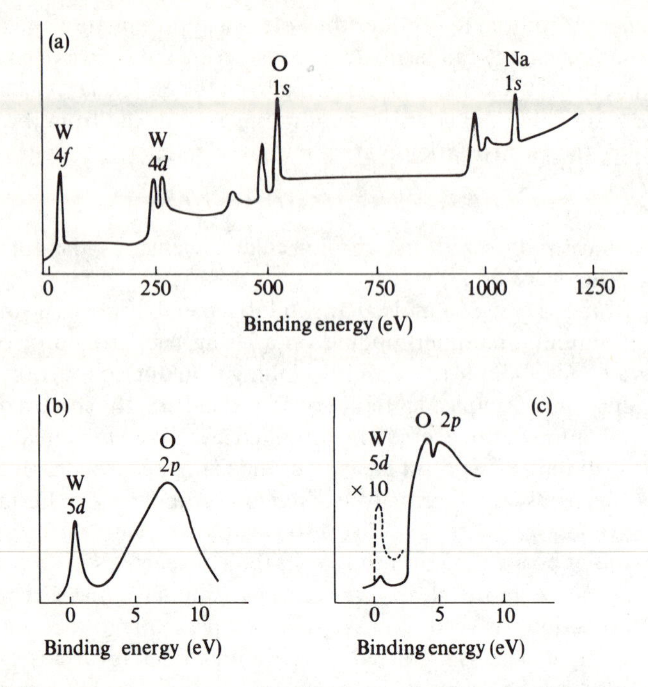

Fig. 2.5 Photoelectron spectra of sodium tungsten bronze, $Na_{0.7}WO_3$. (a) Wide-scan X-ray spectrum showing core levels; (b) X-ray spectrum of valence region showing the W 5*d* and O 2*p* bands; (c) UV spectrum of valence region.

and the dominant features in XPS are the core levels of sodium (1*s*), tungsten (4*f*, 4*d*) and oxygen (1*s*). Extra peaks in the spectrum are caused by Auger transitions, in which atoms that have lost an electron by ionization of a core level lose energy by emitting further electrons. XPS confirms that core levels are relatively unaffected by chemical bonding, and remain as sharp in molecules and solids as they are in free atoms. Small shifts in binding energy are observed, and these can give some information about the oxidation states and charges on the different atoms. However, for various reasons such information is not very reliable in compounds with complex electronic structure, and we shall not discuss core binding energies further.

Information about valence electrons can be obtained from XPS (as shown in Fig. 2.5(b)), but the UV technique is generally more powerful for valence levels, because the exciting line has a much lower energy spread, so that better resolution may be obtained (see Fig. 2.5(c)). The valence region spectra show two bands. The one at lower binding energy, corresponding to the top occupied level, decreases in relative intensity as the photon energy is changed from 1253 to 21 eV. A similar change in relative band intensities is generally found for *d* orbitals in transition-metal compounds, and results from the change in ionization cross-sections as the exciting photon energy is varied. It will be seen later that the top-occupied level in the tungsten bronze is expected to be largely made up of tungsten 5*d* orbitals. This compound has metallic properties due to the electrons in the *d* band. The band at higher binding energy is composed mostly of oxygen 2*p* valence orbitals, and PE signals at a similar energy to this are shown by all metal oxides.

2.2.3 Inverse PES

The PES technique depends on ionizing electrons from solids, and is therefore only capable of showing occupied orbitals. Similar direct information about empty levels can be obtained with a recently developed technique where the PES experiment is reversed. The solid is exposed to a beam of electrons of known energy. Some of the electrons enter the solid, and undergo transitions to empty states in the conduction band, emitting a photon. From the known energies of the electron in its initial state, and of the emitted photon, it is possible to deduce the conduction band energy where the electron ends up. This illustrated schematically in Fig. 2.1(c). The radiation caused by charged particles as they interact with matter is known as *Bremsstrahlung* (literally 'braking radiation', i.e. radiation emitted on deceleration of a fast particle), and the experiment is sometimes known as **Bremsstrahlung spectroscopy**. However, it is important to recognize that it is the direct inverse of the photoelectron experiment, and so it is more appropriate to call it **inverse photoelectron spectroscopy**.

It is particularly interesting to combine the two techniques, and so to give a complete picture of the bands, including both empty and filled states. This is illustrated in Fig. 2.6 for some metals from the first transition series. PES and inverse PES have been drawn on the same energy scale, with the spectra aligned at the Fermi level. States below the Fermi level are seen in PE spectra, and those above the Fermi level in inverse PES. The peak in the middle of these combined spectra shows the band made up of the transition-metal 3*d* orbitals. It can be seen clearly that as the series is traversed from iron to copper, the *d* band progressively fills, and becomes narrower. In nickel the band is not quite full, and some 3*d* levels still appear above the Fermi level in the inverse PE spectrum. In copper however, the *d* band is full, and appears below the Fermi level in the PE spectrum only. The region of the Fermi level now shows a much lower density of states. These features of the electronic structure of transition metals will be discussed in more detail in Chapter 3.

2.3 X-ray spectroscopy

2.3.1 General principles

The core orbitals of an atom are highly contracted, and do not overlap significantly with other atoms in a molecule or solid. We would expect

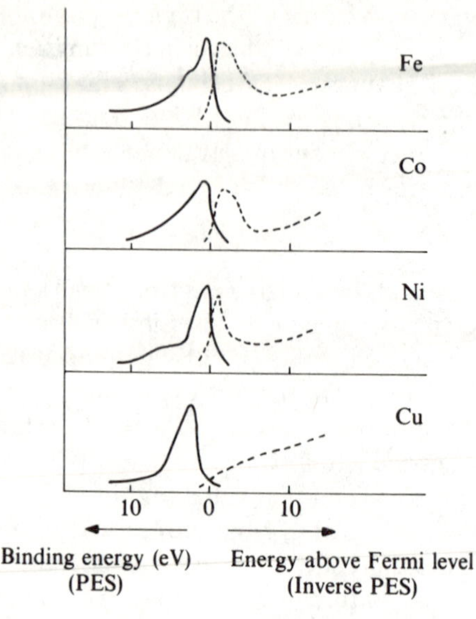

Fig. 2.6 Photoelectron (———) and inverse photoelectron (·········) spectra of the transition elements Fe, Co, Ni, Cu, showing the 3*d* band. (After S. Hüfner and G. K. Wertheim, *Phys. Letters*, **47A** (1974), 349; R. R. Turtle and R. J. Liefeld, *Phys. Rev.*, **B7** (1973), 3411.)

therefore that these orbitals would not be influenced much by the formation of chemical bonds, but would retain their atomic identity. This is confirmed by studies such as the PES experiments just described, where it is found that core levels give sharp lines, unlike the broad bands of valence orbitals. The sharp, atomic character of core orbitals can be made use of in X-ray spectroscopies, where the high photon energies are appropriate to transitions between core levels, or between core and valence orbitals. There are two possible techniques, illustrated in Fig. 2.1. In X-ray absorption spectroscopy (XAS), an electron is excited from a filled core orbital into an empty level in the conduction band of the solid. In X-ray emission (XES), a hole is first generated in a core orbital, by electron or X-ray bombardment. Then the emission is observed resulting from an electron in transition from an occupied valence level to the core vacancy. Since the core orbital corresponds to a sharp energy level, the structure and width of the signals observed in X-ray spectra should reflect the density of states in the conduction or valence band.

For elements beyond the first row of the Periodic Table, different core levels are available for the X-ray emission or absorption experiments, and these may be used to give information not easily obtainable by PES. Because of the atomic nature of the core levels, we can use the selection rules appropriate to conventional atomic spectroscopy, and especially the rule $\Delta l = \pm 1$ for the angular momentum quantum number l. The allowed transitions are from s to p orbitals, from p to s or d, and so on. Knowing the angular momentum of the core orbital involved in the transition, it is therefore possible to deduce that of the atomic orbitals which make up the different valence levels. The nomenclature in X-ray spectroscopy is unfortunately rather confusing for chemists, since spectroscopists use the notation K, L, M, etc. to denote the principal quantum shell of the core level involved, with subscripts to indicate whether it is an s, p, or d orbital.

2.3.2 Applications of X-ray emission spectroscopy (XES)

An example of the information which can be obtained from XES is shown in the two spectra of elemental silicon in Fig. 2.7. The silicon K spectrum is generated by a vacancy in a 1s orbital, so that only transitions from silicon p orbitals are allowed. In the silicon L_{23} spectrum, transitions are to the 2p core levels, showing mostly valence s levels. It can be seen that the occupied valence band in silicon is about 15 eV broad. The valence atomic orbitals available to silicon are the 3s and 3p, and both of them clearly contribute to the band. However, as might be expected, the lower energy 3s orbitals contribute relatively more to the bottom of the valence band, and the higher 3p to the top. In Chapter 3, the bonding in tetrahedral solids such as silicon will be discussed in more detail.

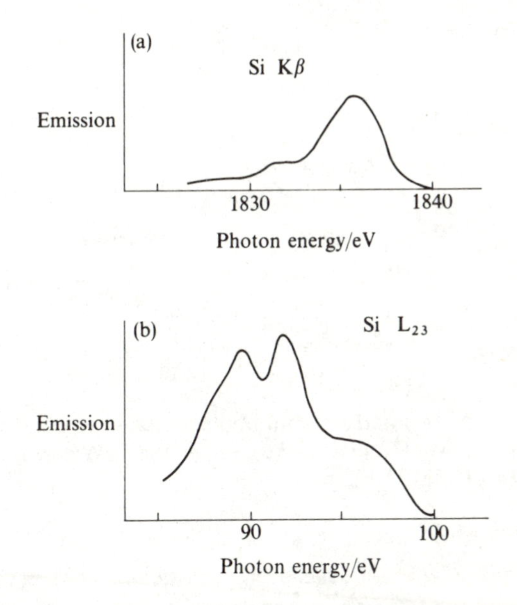

Fig. 2.7 X-ray emission spectra of silicon. (a) Silicon K spectrum, showing Si 3p contribution to filled levels; (b) silicon L_{23} spectrum, showing silicon 3s contribution. (After K. Lauger, *J. Phys. Chem. Solids*, **32** (1971), 609.)

Interesting results have been obtained from the XES of layer compounds such as graphite. The orbital selection rules for dipole transitions predict angular variations in the spectra of solids that are not isotropic. For example, in the carbon K-shell spectrum, a transition from a C $2p_z$ orbital should show maximum X-ray emission in the x–y plane, and no intensity along the z-axis. The spectra of Fig. 2.8 show the carbon K emission from graphite measured with X-rays coming to 10° and 80° from the c-axis (which is normal to the plane of the graphite layers). Emission from the $p\pi$ band, composed of orbitals parallel to the axis, is much less in the 10° spectrum, whereas the signals from the sigma framework band, made up from $2s$ (not seen) and $2p$ orbitals in the plane, are nearly the same in both spectra. Thus it is possible to separate the σ and π contributions. It is clear that the $p\pi$ bonding orbitals form the top-filled levels (giving the highest photon energies in emission), but that the lower part of this band overlaps in energy with the σ levels.

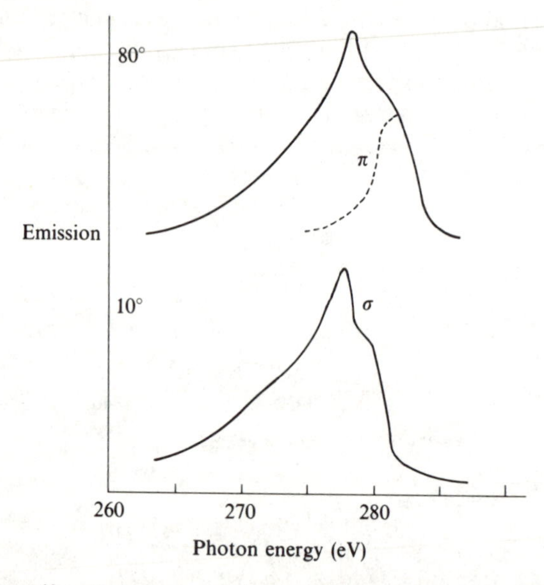

Fig. 2.8 Carbon K emission spectra from graphite, (a) at 10°, and (b) at 80° to the c-axis. The separate σ and π contributions to the filled bands deduced from these spectra are shown. (After H. R. Beyreuther and G. Wiech, *Ber. Bunsenges., Phys. Chem.*, **79** (1975), 1082.)

2.3.3 X-ray absorption spectroscopy (XAS)

The X-ray absorption technique has been exploited rather less than XES. This is partly because information about empty levels is of less direct use than that on filled orbitals, but also because it has been difficult, until the recent use of very expensive synchrotron radiation, to find suitable X-ray sources of variable

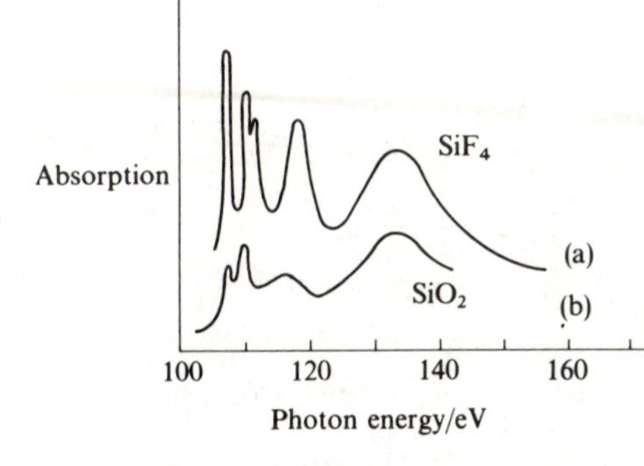

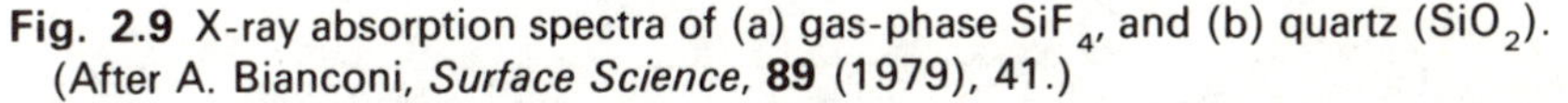

Fig. 2.9 X-ray absorption spectra of (a) gas-phase SiF_4, and (b) quartz (SiO_2). (After A. Bianconi, *Surface Science*, **89** (1979), 41.)

energy. Figure 2.9 shows an example of the kind of information that can be obtained by XAS. The silicon L_{23} XAS of quartz, SiO_2, is seen to be remarkably similar to that of the molecule SiF_4 measured in the gas phase. The absorption spectra result from the transition of electrons from the 2*p* core orbitals of silicon into empty levels. In both compounds, silicon is surrounded tetrahedrally by a more electronegative element, and it seems that this gives rise to conduction band levels in quartz that are very similar to the unfilled antibonding orbitals in SiF_4 molecules. This illustrates quite clearly how the electronic structure of a solid, even when it forms an extended covalent lattice like that of quartz, is largely dominated by the local bonding interactions between near-neighbour atoms.

Electrons at very high energies in a conduction band behave as if they are nearly free. Thus electronic transitions to such high energy states might be expected to be rather uninteresting. In fact XAS measurements do show structure extending to energies of some hundreds of electron volts above the threshold. This is known as the **extended X-ray absorption fine structure** (EXAFS). The EXAFS structure is caused by back-scattering of electrons from atoms surrounding the one being excited. Interference effects occur between the out-going electron wave and the back-scattered waves. At some energies, constructive interference increases the amplitude at the atom which is absorbing radiation, thus increasing the absorption. At other energies, the interference is destructive, and the absorption is decreased. The interference depends on the wavelength of the electron states, and on the path length to and from the surrounding atoms. Since the wavelength can be determined from the kinetic energy in the conduction band, the positions of peaks in the EXAFS spectrum give structural information about the nearby atoms. An important feature of the EXAFS technique is that it is only sensitive to local coordination, and does not depend on long-range crystalline order as do conventional diffraction techniques.

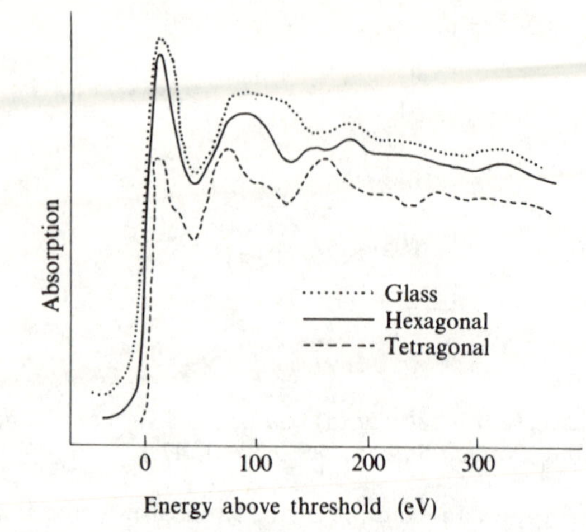

Fig. 2.10 Oxygen K EXAFS spectra of glassy GeO_2, compared with hexagonal and tetragonal crystal modifications. (After W. F. Nelson, I. Siegel, and R. W. Wagner, *Phys. Rev.*, **127** (1962), 2025.)

An example of the use of EXAFS is shown in Fig. 2.10. The spectra shown are of germanium dioxide, GeO_2, in two different crystal forms, and as a glass. The peaks in spectrum of the glass occur at energies very similar to those of the hexagonal (quartz) crystal modification, and different from tetragonal (rutile) form. This shows that glassy GeO_2 probably has tetrahedral coordination of oxygen atoms around germanium, similar to that in the quartz form, and not octahedral coordination as in the rutile structure. Changes in bond lengths can be found from more detailed analysis. This kind of information about the structure of amorphous solids is difficult to obtain from more conventional diffraction techniques.

Both XES and XAS have a problem which is shared to a lesser extent by other forms of spectroscopy. The electronic levels are never observed in the ground state of the solid, but always during some disturbance caused by an electronic transition. In X-ray spectroscopies the disturbance involves a hole in a core orbital, created either during the X-ray absorption process, or previously in the emission experiment. Making a core hole decreases the screening of the nuclear charge experienced by valence electrons. In some cases, such as in the examples discussed here, this produces only a minor perturbation on the electronic structure, and does not complicate the interpretation of the results. Sometimes however, the core hole has a very serious effect on the valence electrons, and consequently it is much more difficult to obtain simple information about electronic structure of the unperturbed solid.

2.4 Optical properties

The interaction of a solid with visible and UV radiation involves the excitation of valence electrons, for example from filled to empty bands, or within partially filled bands. At a simple level, the visible appearance of a solid can give many clues to its electronic structure. More quantitative measurements of visible and UV spectra give important information about the valence and conduction bands. Unfortunately, we shall see that the optical properties of solids can be quite complicated, and involve some concepts which are unfamiliar to chemists. (The material in Sections 2.4.2 and 2.4.3 is not essential for understanding most of the rest of the book, and could be missed out on a first reading.)

2.4.1 Band gaps and excitons

The simplest picture, discussed in Chapter 1, suggests that in a non-metallic solid, where there is an energy gap between the filled valence band and the empty conduction band, only photons with energy greater than the gap will be absorbed. Photons of longer wavelength will pass through, having insufficient energy to excite an electron. Thus measuring the threshold for optical absorption is a useful simple method for estimating the band gap. Figure 2.11 shows absorption spectra of: (a) a typical insulator, sodium chloride, where the onset of absorption is around 7.5 eV in the far UV, and (b) a semiconductor, gallium arsenide, where the threshold is 1.5 eV, just at the borderline of the visible spectrum with the infra-red.

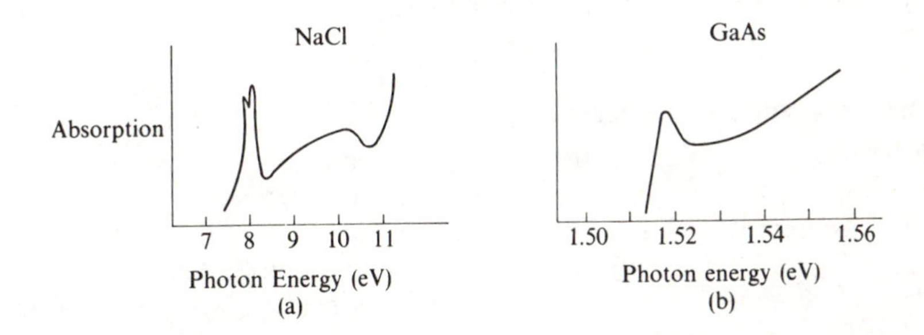

Fig. 2.11 Absorption spectra of (a) NaCl measured at 80 K; (b) GaAs measured at 4.2 K. (After J. E. Eby, K. J. Teegarden, and D. B. Dutton, *Phys. Rev.*, **116** (1959), 1099; M. D. Sturge, *Phys. Rev.*, **127** (1962), 768.)

Most of the data on band gaps in solids, which we shall refer to in future chapters, come from absorption spectra such as in Fig. 2.11. Unfortunately however, the process of optical absorption can be a little more complicated than described above. Both spectra shown in Fig. 2.11 have peaks at the absorption threshold. These peaks are caused by an **exciton**, which is an excited state of the solid where the electron does not escape fully into the conduction band, but remains trapped in the electrostatic potential of the hole left behind in the valence band. It is only at slightly higher energies that the electron and hole become free, and the absorption corresponds to the true band gap. Not all solids show such excitons, and in many of those which do, such as gallium arsenide, they are seen only at low temperatures, where the broadening effect of lattice vibrations is reduced. In Chapter 7, we shall look in more detail at the factors which influence the binding of an electron and a hole. We shall also see in that chapter that optical absorption may be caused by defects and impurities in the solid.

When excitons or defect bands are present in the absorption spectrum, the estimation of the true band gap can be quite difficult. Sometimes it is possible to fit a theoretically derived curve to the exciton peak, and subtract it, thus revealing the true inter-band absorption spectrum. It is also possible to utilize the appearance of **photoconductivity**, which occurs when unbound electrons and holes are produced by absorption of light. In the exciton, the electron and hole are held together by their mutual electrostatic attraction, and so form an electrically neutral entity that cannot give rise to conductivity. The photon energy corresponding to the onset of photoconductivity gives a more reliable measure of the energy gap than is possible from absorption alone. In semiconductors such as gallium arsenide, the thresholds for absorption and photoconduction are almost identical, showing that in these cases the exciton peak is virtually on top of the band gap threshold. In alkali halides, on the other hand, the exciton is bound quite strongly, and the onset of optical absorption may be 1–2 eV below the true band gap.

2.4.2 Absorption and reflectivity

The simple theory of absorption spectroscopy, applicable to species in dilute conditions in the gas phase or in solution, considers only absorption and transmission of light. Solids on the other hand may show strong **reflectivity** of light; indeed it is often only possible to study the spectrum of a solid by light reflected, rather than transmitted. Since reflectivity also plays a major part in determining the appearance of a solid, it is interesting to look at the factors which influence it.

The behaviour of light when it passes from air (or vacuum) into a solid is determined by the refractive index, n, of the solid. Normally this is taken as a real number, but at wavelengths where some absorption occurs, it is convenient to represent the absorbance as an imaginary contribution to n. Thus we write:

$$n = n' - i n'' \tag{2.3}$$

where n' represents the normal refraction, and n'' the absorption strength. The proportion of light reflected back from the solid is given by the formula:

$$R = \frac{(1-n')^2 + n''^2}{(1+n')^2 + n''^2}. \tag{2.4}$$

Thus it can be seen that reflection occurs, not only when the refractive index n' is different from one, but also in regions of absorption, where n'' is non-zero. In fact it appears paradoxically that very strong absorption by the solid, when n'' is large, manifests itself as high reflectivity. This happens in many solids, either well above the band gap in a non-metal, or down to very low photon energies in a metal where there is no gap. A semiconductor such as silicon, with a gap of 1.1 eV corresponding to the near infra-red, appears quite reflective and metallic to visible light.

A more detailed picture of the optical properties of a solid depends on knowing how the complex refractive index varies with the photon energy. In a non-magnetic solid, the refractive index is the square-root of the relative dielectric constant, ε_r. The variation of ε_r with photon energy or frequency is known as the **dielectric function**, and is an important property of a solid. Dielectric functions may be deduced experimentally from reflectance or absorption measurements, and often vary in a complicated way with energy. However, the important features may be shown in the following simple model, which is intended to mimic the behaviour of a solid with one optical absorption band. The absorbing group may be represented classically by a charged electron vibrating on a spring (see Fig. 2.12(a)). We suppose that the force constant of the spring gives rise to a vibrational frequency ω_0, and that interaction with the surroundings damps the oscillation with a characteristic decay time τ. A dilute collection of N such oscillating electrons in solution or in the gas phase, would have an absorption coefficient given by:

$$\alpha(\omega) = \frac{(Ne^2/\varepsilon_0 m)(\omega/\tau)}{(\omega_0^2 - \omega^2)^2 + (\omega/\tau)^2} \tag{2.5}$$

As Fig. 2.12(b) shows, this corresponds to an absorption line centred on the fundamental frequency, ω_0. The width of the line is controlled by the damping, and is around $1/\tau$.

We now consider an array of oscillators in a solid, with N per unit volume. Each oscillating electron experiences electric fields from the other electrons, as well as any field applied externally to the solid. An electrostatic calculation which takes account of this mutual interaction leads to the following expression for the dielectric function $\varepsilon(\omega)$, of the solid:

$$\varepsilon(\omega) = 1 + \frac{\omega_p^2}{(\omega_0^2 - \omega_p^2/3) - \omega^2 + i\omega/\tau} \tag{2.6}$$

where

$$\omega_p^2 = (Ne^2/\varepsilon_0 m). \tag{2.7}$$

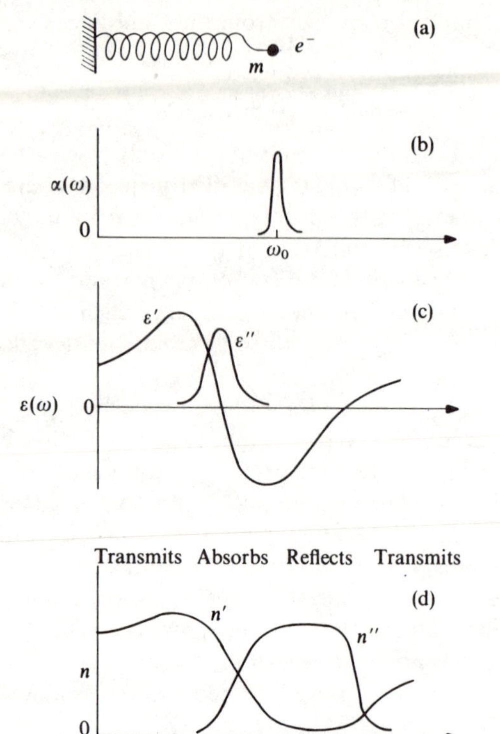

Fig. 2.12 Dielectric and optical properties of a model solid. (a) Electron on a spring, to simulate a single absorption band; (b) absorption spectrum expected for (a) in a dilute system; (c) real (ε') and imaginary (ε'') parts of the dielectric function resulting from an array of (a) in a solid; (d) real (n') and imaginary (n'') parts of the refractive index. Also indicated are the principal regions of transmission, absorption, and reflection from the solid.

As will be seen in Figs. 2.12(c) and (d), this dielectric function has real and imaginary parts ε' and ε'' respectively, and gives rise to the kind of complex refractive index discussed above. Figure 2.12 also indicates roughly the main regions of optical transmission, absorption and reflection, although these merge into each other, and there will be some degree of reflectance across the whole spectral range. The region of strongest absorption corresponds to the peak in the imaginary part of the dielectric function, ε'', and it appears that this has been shifted down in energy from the value ω_0 in the isolated system. The shift results from the interaction between the oscillators in the solid, through the electric field which each one produces.

The model described is clearly a very idealized one, and real solids have a range, or indeed several ranges, of absorption frequency. Nevertheless the picture which emerges is quite useful. It can be seen in Fig. 2.12 that a region of transmission at long wavelengths (low frequencies) is succeeded by an absorption band, corresponding to the band gap or exciton discussed in the previous section. At higher frequencies, there is a range of high reflectance, before transmission is possible again. The appearance of a solid depends on the degree of absorption and reflectance in the visible region of the spectrum, where photon energies are in the range 1.5–3 eV. Generally speaking, solids with an energy gap in this range will appear coloured, first yellow, then deeper orange and red, as the band gap moves down through the visible. With a band gap below 1.5 eV, in the infra-red, the solid may appear dark in colour or shiny metallic, depending on the reflectivity. An absorption in the near infra-red tailing into the visible may lead to a blue appearance, as happens for instance with the low-sodium tungsten bronze, $Na_{0.3}WO_3$.

When the band gap is greater than 3 eV, no absorption occurs in the visible, and a good quality crystal will be transparent. Light is scattered however, at crystal flaws and at the crystallite surfaces in a powdered sample, and this leads to a white appearance. The best white pigments are those which have no absorption in the visible, but high reflectivity. This means that the band gap must be at least 3 eV, and the refractive index (and hence dielectric constant) as high as possible. Figure 2.12 shows that the maximum refractive index is found just below the threshold for absorption. Thus for maximum reflectivity of visible light, the absorption edge should be as close to the visible as possible, that is just over 3 eV. This is the case for rutile, TiO_2, which is also cheap, non-toxic and chemically stable, and is the most widely used white pigment.

Solid compounds generally show vibrational spectra in the infra-red region, and the model described above may also be used to treat these. Now the mass and charge in equation 2.7 must be replaced by the reduced vibrational mass and an effective charge carried by the vibrating atoms. This modified formula is discussed in Section 3.2.3 of Chapter 3. As with optical spectra, there is a region of high infra-red reflectivity corresponding to each vibrational mode. For reasons concerned with the detailed atomic motions, the frequencies corresponding to the lower and upper limits of the reflecting region are known as the **transverse** and **longitudinal** vibrational frequencies, respectively.

2.4.3 Metals—the plasma frequency

A metal has no energy gap above the top-filled level, and thus the minimum absorption frequency is zero. From Fig. 2.12, the region of high reflectivity should now extend down to zero frequency. A simple dielectric function of a metal may be derived by putting the absorption threshold to zero, and ignoring the imaginary part in the denominator of equation 2.6:

$$\varepsilon(\omega) = 1 - \omega_p^2/\omega^2 \tag{2.8}$$

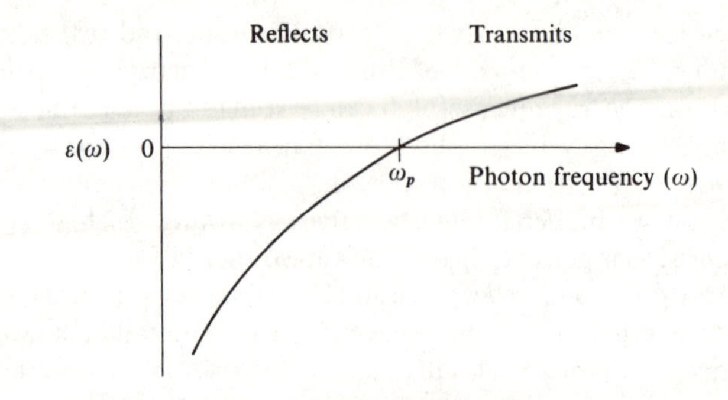

Fig. 2.13 Dielectric function of a metal, showing the plasma frequency (ω_p) and the regions of reflection and transmission of light.

where ω_p is given as before by equation 2.7, and is called the **plasma frequency**. The dielectric function of a metal is shown in Fig. 2.13. Below the plasma frequency the dielectric constant is negative, and so the refractive index must be purely imaginary, that is n' is zero. Equation 2.4 then shows that the reflectivity is unity. Above the plasma frequency, however, the metal is rather unexpectedly predicted to become transparent. This happens with sodium, for example, at wavelengths in the UV shorter than 209 nm, corresponding to an energy $\hbar\omega_p$ of 5.8 eV, in good agreement with the predictions of equation 2.7. The plasma frequency depends on the electron density N, and in most metallic elements is in the far UV. Thus the typical appearance of a metal results from its high reflectivity in the visible region. There are many complications in real metals, which lead to some absorption below the plasma frequency. The colour of a metal arises because the reflectivity is less than perfect, and varies through the visible spectrum. In some metallic compounds however, the density of conduction electrons may be much lower than in metallic elements, and plasma frequencies may be in the visible spectrum or even in the infra-red. For example tin dioxide, SnO_2, becomes metallic when it is doped with around 1 per cent of antimony. The low concentration of conduction electrons gives a plasma frequency in the infra-red, and the solid is consequently transparent to visible light.

In Section 2.4.1 we described optical absorption in terms of the excitation of individual electrons between different energy levels. This picture is certainly not applicable to the optical properties of a metal, which are dominated by the very strong interaction of conduction electrons with electromagnetic radiation. These electrons also interact strongly with each other, through the electrostatic repulsion of the negative charges. In fact, the plasma frequency given by equation 2.7 corresponds to an unusual kind of **collective excitation** in which all the conduction electrons move in phase. This can be imagined as a kind of compression wave, rather like a sound wave in a gas. The frequency of these oscillations is determined by the electrostatic repulsion between electrons, which explains the appearance of the density N and the electronic charge e in the equation 2.7. As with atomic vibrations, the plasma oscillations are quantized in energy units of $\hbar\omega_p$, and known as **plasmons**. Plasmons cannot be excited directly by optical absorption, although their energy can be measured from the reflectance edge in a metal. However, plasmons in metals can be excited very efficiently by a beam of charged particles such as electrons, and appear strongly in **electron energy loss spectroscopy**. In this experiment, a beam of electrons is either fired through a thin film of solid, or reflected from its surface. The energy spectrum of the scattered electrons is measured, and loss peaks, corresponding to electrons which have produce some vibrational or electronic excitation, can be observed.

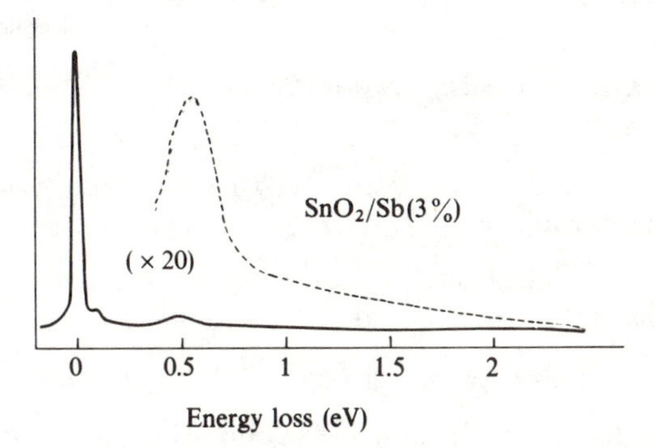

Fig. 2.14 Electron energy loss spectrum of 25 eV beam reflected from tin dioxide (SnO_2) doped with 3% antimony. The plasmon is the amplified peak at about 0.5 eV loss energy. (After P. A. Cox *et al.*, *Solid State Commun.*, **44** (1982), 837.)

An example of a high-resolution electron energy loss spectrum, from a sample of tin dioxide doped with antimony as mentioned above, is shown in Fig. 2.14. The low-energy electron beam (25 eV) was scattered off the surface of the sample. The strongest peak corresponds to electrons scattered elastically, with no energy loss, and close to this is a weak feature from the vibrational excitation. The peak at 0.5 eV energy loss is the plasmon, with energy $\hbar\omega_p$. The electron energy-loss experiment gives a much more direct measure of the plasma frequency than is possible from optical spectroscopy. This can provide useful information about the conduction electrons. For example, in the doped tin dioxide of Fig. 2.14, it appears that each antimony atom donates one electron to the conduction band.

Further reading

The general principles of photoelectron spectroscopy, with some applications to solids, are described in:

D. Briggs (ed.) (1977). *Handbook of X-ray and ultra-violet photoelectron spectroscopy*. Heyden.

Reviews of the application of spectroscopic techniques to solids can be found in:

P. Day (ed.) (1981). *Emission and scattering techniques*. D. Reidel.

Of particular relevance to the material of this chapter are the articles by Cox (photoelectron spectroscopy), Wiech (X-ray spectroscopy), and Baer (inverse photoelectron, or Bremsstrahlung, spectroscopy).

The articles in the following volume give a much more detailed account of the applications of photoelectron spectroscopy to different types of solid:

L. Ley and M. Cardona (eds.) (1979). *Photoemission in solids. II. Topics in Applied Physics*, vol. **27**. Springer-Verlag.

Another volume concerned with the spectroscopy of solids which includes an article by Urch on X-ray spectroscopy is:

F. J. Berry and D. J. Vaughan (1985). *Chemical bonding and spectroscopy in Mineral Chemistry*. Chapman and Hall.

A general account of the optical properties of solids can be found in:

M. V. Klein (1970). *Optics*, Section 11.2. John Wiley and Sons.

An elementary account of plasma excitations and the optical properties of metals is:

C. Kittel (1976). *Introduction to solid state physics*, (5th edn), Chapter 10. John Wiley and Sons.

3 Electronic energy levels and chemical bonding

The most important features of the electronic structure of a solid are the energies and widths of the various bands, the energy gaps between them, and the number of electrons which occupy them. We have seen in the previous chapter that this information can be obtained by various types of spectroscopic measurement. In this chapter, we shall consider in more detail how the electronic energy levels of simple solids are related to chemical models of bonding.

We shall not discuss **molecular solids** any further at this point. The weak van der Waals' forces between molecules have rather little influence on the major features of electronic structure discussed in this chapter. The interaction between molecules can be more significant in their excited states, and gives rise to some effects mentioned in Chapter 7. There is also a small class of molecular solids where stronger intermolecular interactions give rise to remarkable properties, including metallic conductivity. The electronic properties of these so-called 'molecular metals' depend on rather subtle features, however, and their treatment is postponed until Chapter 6. The first class of solids that we shall discuss in detail here is that described by the **ionic model**, where bonding is accompanied by an electron transfer between one atom and another.

3.1 Ionic solids

The ionic model can be used to predict the heats of formation and many of the chemical and physical properties of compounds such as simple halides and oxides, where an electropositive metal is combined with a very electronegative non-metal. According to the ionic picture, the bonding is provided by the transfer of electrons from the metal to the electronegative partner. The ionic model predicts that the valence band, which forms the top-filled level in the solid, should be made up of the top-occupied orbitals of the non-metal anion; the bottom empty level, the conduction band, is correspondingly formed from the lowest empty orbitals of the metal cation. It is easy to see why simple ionic compounds are good insulators. The fact that each ion has a closed shell electron configuration ensures that all bands are either completely full, or empty. The situation is different in transition-metal compounds, where there may be a partially occupied shell of valence *d* electrons left on the metal ion. Some examples of transition-metal compounds are considered later in the chapter.

The alkali halides form the series of compounds where the ionic model is best obeyed. We shall start with a typical member of this series, and look in detail at the factors which determine the energies of filled and empty levels.

3.1.1 Sodium chloride—a typical example

In the simple chemical picture, formation of sodium chloride from its elements involves the transfer of an electron from sodium (which is left in the crystal as the cation Na^+) to the chlorine (left therefore as an anion Cl^-). The top-filled level is expected to be the chlorine 3*p* orbital into which an electron is placed in making the ion; and the bottom empty orbital, the sodium 3*s* from which an electron has been lost. The first step in calculation of the energy gap (as shown in Fig. 3.1) would be to consider the energy of transferring the electron back

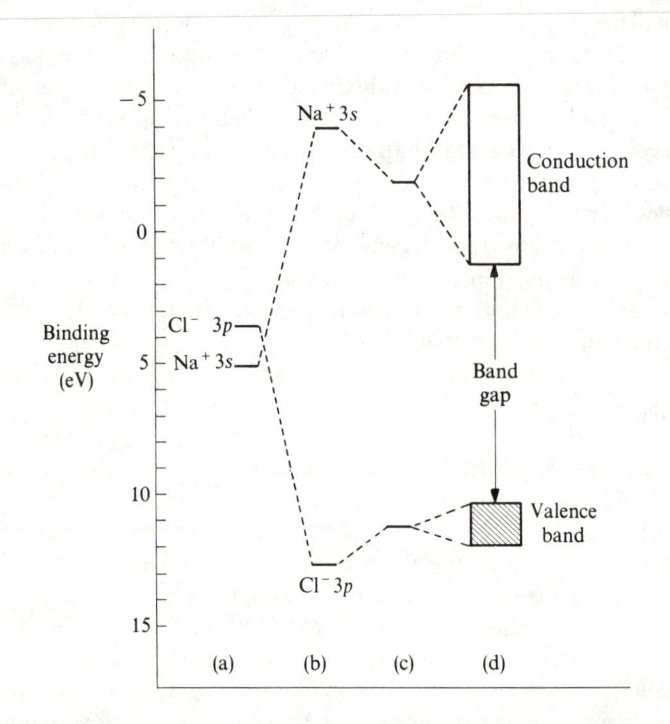

Fig. 3.1 Derivation of the valence and conduction band energies of sodium chloride. (a) free ions; (b) ions in the Madelung potential of the lattice; (c) correction for electrostatic polarization on removing or adding an electron; (d) inclusion of bandwidths arising from orbital overlap.

again, that is, the energy of the process:

$$Na^+ + Cl^- = Na + Cl.$$

The energy of this reaction in the gas phase is simply the combination of that required to remove an electron from Cl^-—which is the same as the electron affinity A of a chlorine atom—and the energy involved in placing the electron on Na^+, which is minus the ionization potential I of sodium:

$$E_g = A - I. \tag{3.1}$$

The result for sodium chloride is -1.5 eV. Negative values are also found for nearly all other ionic solids, showing that in the gas phase the ionic configuration is less stable than that of neutral atoms.

The first, and most important, correction to be made to the initial estimate, is to recognize that the ions in the solid experience a strong electrostatic potential from the surrounding ions of opposite charge. We have seen in Section 1.2.2 (p. 4) that the long-range nature of the Coulomb force means that we must take into account not just the near-neighbour ions, but also the potential from more distant shells of ions in the crystal lattice. Just as in calculation of the lattice energy in the simple ionic model, the potential from all the successive shells of ions involves the Madelung constant A_M of the lattice. The resulting potential is therefore called the **Madelung potential,** V_M, and is related to the interionic distance r by:

$$V_M = A_M ze/(4\pi\varepsilon_0 r) \tag{3.2}$$

The value of V_M is 9 V for sodium chloride, and can be substantially more when there are ions such as O^{2-} with multiple charges. The Madelung potential is positive (attractive for an electron) at the anion site, and gives a value of:

$$E_B = A + eV_M \tag{3.3}$$

for the binding energy of an electron in the $3p$ valence level of Cl^- (see Fig. 3.1(b)). The same term contributes as a repulsive energy on the cation site, and so the estimate of the band gap including the Madelung potential is now:

$$E_g = A - I + 2eV_M \tag{3.4}$$

This gives 17 eV, which is still a long way from the experimental value of 9 eV, but now at least has the correct sign. The importance of the Madelung potential in stabilizing the ionic charge distribution is clear.

The description of the ions in the crystal lattice is still too simplistic, and there are two further factors that we need to consider. We noted earlier (Section 2.2.2, p. 29) in the comparison of gas-phase and solid-state photoelectron spectra, that the binding energy of an electron in the solid is lowered by the polarization of the surrounding crystal when an electron is removed. Equation 2.2 gives an estimate for the polarization energy, making the approximation that the solid is a continuum with a certain dielectric constant. This equation predicts a polarization energy of 1.45 eV, when a hole is made in a chlorine $3p$ orbital in NaCl. More detailed calculations have been performed, in which the electric fields at nearby ions, and the polarization that they produce, are calculated explicitly. Only the more distant ions in the lattice are approximated by a polarizable continuum. Such calculations for NaCl give only a slightly different value of 1.55 eV. Because the polarization energy lowers the binding energy of an electron, it is represented in Fig. 3.1(c) by raising the energy of the chlorine $3p$ orbitals. The polarization also gives an energy lowering when an electron is placed on a Na $3s$ orbital, and the magnitude of this is somewhat larger, about 2.5 eV, because the chloride ions (which are near-neigbours to sodium) are more polarizable than the sodium neighbours round a chloride ion.

The final effect which must be allowed for is the overlap of orbitals between neighbouring ions to form bands. Measurements by photoelectron spectroscopy (PES) show that the valence band in NaCl is quite narrow, less than 2 eV in width. Indeed, the measured binding energy for electrons in the valence band is 10.8 eV, which agrees well with the value of 11 eV calculated for the $3p$ level without the bandwidth included. However, the estimate of the band gap without allowing for bandwidths is still too large by about 4 eV. It appears that the conduction band formed from the sodium $3s$ orbitals is quite broad, probably at least 5 eV. The difference in width between the valence and conduction bands is not surprising, as the chlorine $3p$ orbitals are fairly contracted, and so do not overlap much with each other in NaCl, whereas the Na $3s$ orbitals are much more diffuse, and overlap extensively with neighbouring ions.

3.1.2 General trends

The calculation given above for sodium chloride confirms the simple chemical picture of an ionic solid: the valence band, being the top-filled level, is composed of anion atomic orbitals, and the conduction band, (the bottom empty level), of cation orbitals. However, we saw that a quantitative estimate of the band gap is quite difficult even in this simple case, because of the variety of effects that need to be included. Similarly detailed calculations can be made for other ionic compounds, but it is more useful at this point to give a qualitative discussion of the general trends.

Our detailed discussion of NaCl has shown that the single most important term in determining the band gap of an ionic solid is the Madelung potential, V_M, given by equation 3.2. In the alkali halide series, the difference between electron affinity of the anion and the ionization potential of the cation is almost negligible in comparison. The other terms included in Fig. 3.1, that is the polarization and bandwidths, are important but like V_M they decrease with the interionic separation. Taking these factors together, we would predict that the band gap should generally decrease with increasing lattice parameter. Figure 3.2(a) shows a plot of the band gaps for alkali halides against the inverse anion–cation distance. We can see that this does indeed give the correct trend for alkali halides. Lithium fluoride, LiF, has the largest band gap of any solid, and the gaps decrease with increasing size, either of the halide or the alkali metal ion. This is a rather clear indication of the importance of the $1/r$ Madelung term in contributing to the band gap in these compounds.

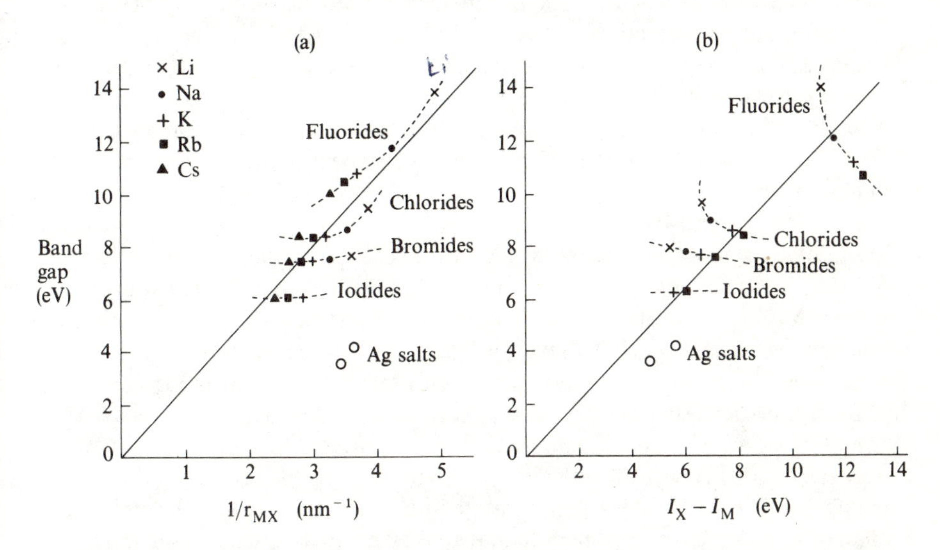

Fig. 3.2 Band gaps for halides MX plotted against: (a) inverse of cation–anion distance; (b) difference of neutral-atom ionization energies $I(X) - I(M)$. (In each plot, the straight line through the origin is drawn as a guide to the major trend, and has no theoretical significance.)

The band gaps of some silver halides are also plotted in Fig. 3.2, and they seem to fall rather far from the values of the alkali halides. Silver is in the post-transition metal Group IB, and as a consequence of the filling of the *d* shell, its 5*s* ionization potential is considerably higher than in the alkali metal series. In ionic terms, it has about the same radius as sodium, but the latter has an ionization energy of 5.1 eV, compared with 7.6 eV for silver. Thus it is no longer a reasonable approximation to ignore the atomic terms, $A - I$ in equation 3.4. It is characteristic of salts of post-transition metals that they have smaller band gaps than the corresponding pre-transition metal compounds. Since the gaps may be brought into the visible region of the spectrum, compounds of post-transition metals are often coloured. For corresponding ionic compounds of the alkali metals and alkaline earths, the gaps are always greater than 3 eV, and absorption only starts in the UV.

The comparison of sodium and silver suggests a correlation of the band gap with atomic energies. Clearly, as the case of NaCl shows, we cannot use the difference between the electron affinity of the halogen and the ionization potential of the metal. However, the result of the calculation in the previous section shows something quite interesting: the difference between electronic levels of atoms and ions in the gas phase is almost cancelled by placing them in the crystal lattice. The final positions of the band edges are not so far away from the *neutral atom* electronic energies, measured as ionization potentials. This cancellation is certainly not exact, but it can be understood by considering the process of forming an anion from a neutral halogen atom, and then putting it into the crystal lattice. The difference between the ionization potential and electron affinity of the halogen is due to the extra electron repulsion present in the anion. In the lattice, this extra repulsion is compensated by the attraction to neighbouring cations.

Figure 3.2(b) shows a plot of the band gap, against the difference of neutral atom ionization energies, for some halides. The correct trend is obtained when halogen is varied, but not the alkali metal. According to the atomic ionization energy, lithium compounds should have the smallest gaps of the alkali halides, whereas we have seen that they have the largest, because of the Madelung potential. The silver compounds, however, are now closer to the trend.

It can be seen from this discussion that the trends in band gaps are complicated by the competition of different effects. Moving down a pre-transition metal group, the most important trend is the increase in ionic radius, giving a decrease in band gap through the Madelung potential term. The change in atomic ionization potential makes a much smaller contribution. Moving down the group of halogens, the two factors—ionization potential and Madelung potential—go together, and there is a clear decrease in band gap. The same is true for anions in other groups: thus sulphides have smaller gaps than oxides, and as a result are more often coloured. In comparing pre- and post-transition elements, the increase of metal ionization potential leads to a decrease of the gap. Indeed, for these compounds the ionic model is less satisfactory, and considerable deviations can be seen in predicted crystal structures and lattice energies. The deviations come from covalent contributions to the bonding, which are a consequence of the relatively smaller energy differences between cation and anion levels, so that more efficient covalent mixing of the orbitals can take place. Compounds with mixed ionic and covalent character are discussed from a slightly different point of view in Section 3.2.2.

It is also interesting to look at trends in the band widths. Valence band widths can be measured by PES, and some examples are shown in Fig. 3.3(a). The fluorine $2p$ valence bandwidth is seen to fall considerably in the alkaline earth difluorides, from BeF_2 to BaF_2. In Fig. 3.3(b) it is shown that the bandwidth of a series of mono- and di-fluorides with different structures

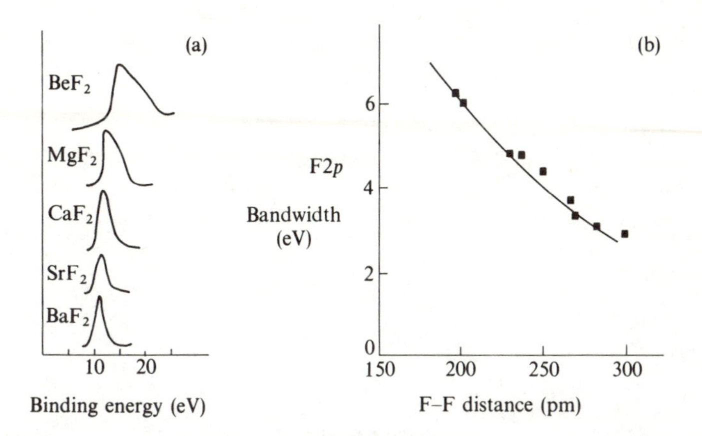

Fig. 3.3 Fluorine $2p$ valence bandwidths of simple fluorides: (a) PES of dihalides of Group IIA elements Ba to Be, showing how the band narrows down the Group; (b) bandwidths of mono- and dihalides plotted against the F–F distance. (After P. T. Poole, S. Szajman, R. G. Leckey, and J. Liesegang, *Phys. Rev.*, **B12** (1975), 5872.)

correlates well with the fluorine–fluorine separation in the crystal lattice. This suggests that the major contribution to the valence bandwidth is direct overlap between anion orbitals, which diminishes with increasing distances. For heavier halides, the overlaps are less and the bandwidths quite narrow (1–2 eV), whereas for oxides (where the anion is more diffuse) valence bandwidths are around 5 eV or more. In the simple ionic model the direct overlap between ions of the same kind is the only possible source of bandwidth, but as the degree of ionic character decreases, valence and conduction bands may also be broadened by the covalent overlap between anions and cations. Covalency is an important source of bandwidth in some of the transition-metal compounds considered in Section 3.4.

3.1.3 Cations with filled *d* and *s* orbitals

The general rule that valence bands are composed of the anion orbitals needs to be qualified sometimes, when the cation also has filled orbitals fairly high in energy. One important example is in ions with the d^{10} electron configuration, following the transition series. In the group IIB elements, zinc, cadmium, and mercury, PES measurements show that the filled *d* shell is quite far down in energy, with much higher binding energies than the halogen *p* levels. In monovalent copper, silver, and gold however, the *d* levels are not so core-like, and are in the same energy region as the anion valence levels. It is not always easy to decide which is the top-filled level, although PES experiments with different photon energies can sometimes help, as illustrated in Fig. 3.4. The valence region of the spectrum of AgI shows two bands, which change in

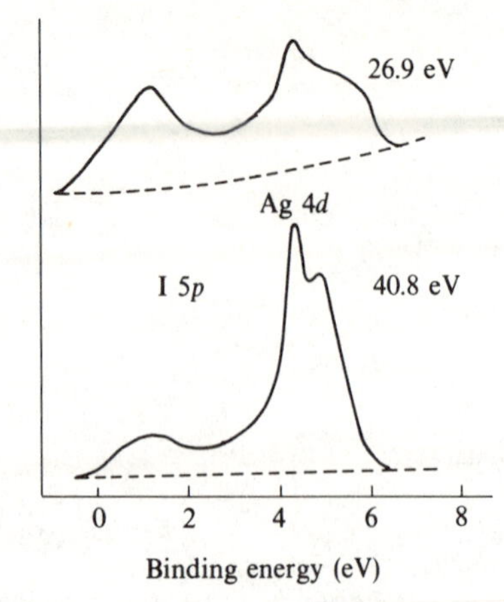

Fig. 3.4 Photoelectron spectra of AgI measured at 26.9 and 40.8 eV photon energies, showing the levels with predominantly I 5*p* and Ag 4*d* composition. The Ag 4*d* ionization cross-section is relatively higher at 40.8 eV. (After A. Goldman, J. Tejeda, N. J. Shevchik, and M. Cardona, *Phys. Rev.*, **B10** (1974), 4388.)

relative intensity as the photon energy is increased from 26.9 to 40.8 eV. At the higher photon energy, it is expected that the metal 4*d* ionization cross-section will be relatively larger. Thus the band at higher binding energy must be assigned to the Ag 4*d* orbitals, and the top of the valence band is mostly iodine 5*p* in character, as in alkali iodides. Sometimes this order is reversed however, as happens with the copper halides CuCl and CuBr. The 3*d* ionization potential in copper is less than that of the 4*d* in silver, and the copper 3*d* orbitals form the top-filled level, above the halogen valence band levels. This interesting difference can be related to the comparative chemistry of copper and silver. It is much easier to oxidize copper to the +2 oxidation state, because of the lower binding energy of its 3*d* orbitals relative to the 4*d* of silver. Spectra of CuI show that the energies of filled metal and anion levels are very close. Solution chemistry suggests that it is easier to remove an electron from an iodide ion than from Cu^+; thus iodide is oxidized by Cu^{2+} to give iodine and CuI.

Similar problems arise with the post-transition metals in lower valence states, where the cation has the s^2 configuration. An example is monovalent thallium, with two 6*s* electrons. Calculations similar to those for NaCl predict that in TlCl the chlorine 3*p* and thallium 6*s* binding energies should be very close—11.2 and 12.2 eV respectively. In fact, the PES of this compound shows a series of bands between 10 and 15 eV binding energy, which are attributed to different mixtures of the orbitals on the two atoms.

The existence of filled cation levels close to the top of the valence band clearly affects the interpretation of the optical absorption spectrum. In most ionic solids, the transition of an electron from the valence band to the conduction band is essentially a charge-transfer from anion to cation. With the s^2 ions, the conduction band is made up of the cation *p* orbitals, and the optical absorption may have a large component of the strongly allowed intra-atomic *s–p* transition.

3.2 Covalent solids

In solids with predominantly covalent bonding, a description of orbitals based on cations and anions is clearly no longer appropriate. The relevant chemical model is now one where atomic orbitals of similar energy overlap and form bonding combinations—which are occupied and make up the valence band—and antibonding ones, which are normally empty and hence form the conduction band. It is simplest to consider first the solids formed by non-metallic elements, where the covalent bonds are non-polar. Subsequently, we shall look at compounds where covalent bonding is combined with some degree of ionic character.

3.2.1 Elemental solids

Many of the non-metallic elements of Groups IV, V and VI in the Periodic Table form solids where all atoms are linked by covalent bonds. The structures of these solids show a coordination number and geometry very similar to that found in molecules where the atom exhibits its normal valency. Thus the Group IV solids are tetrahedrally bonded, those in Group V three-coordinate, and those in Group VI two-coordinate. This strongly suggests that the covalent bonds in the solid are similar to those in small molecules. A carbon–carbon bond in diamond, for example, is very little different from one in a saturated hydrocarbon such as ethane, C_2H_6. The exceptional nature of the first-row elements—carbon which also forms graphite, and nitrogen and oxygen which have diatomic molecular lattices—can be understood in terms of the much greater ability of these elements to form multiple bonds.

The simplest covalent solids to consider—and also the most important in practical terms—are the tetrahedral solids of Group IV. The band gaps, in eV, of these elements in the diamond structure are:

C	Si	Ge	Sn
5.5	1.1	0.7	0.1

In tetrahedral molecules formed by these elements, it is common to think of the bonding in terms of valence *s* and *p* atomic orbitals, used to construct sp^3 hybrids. The MO model of bonding gives a slightly different picture, with delocalized orbitals formed from combinations of the *s* and *p* orbitals of the Group IV atom and atomic orbitals on the surrounding atoms. As we have seen in Section 1.3.2 (p. 14), localized and delocalized views of the electronic structure are really equivalent, being based on different ways of dividing up the total electron distribution. It is possible to use the sp^3 hybrid picture to think about tetrahedral solids. In this view, the valence bands are composed of combinations of hybrids which are bonding between adjacent atoms, and the conduction band of combinations which are antibonding. The decrease in the band gap down the Group can be attributed to the reduction of orbital overlap, which leads to a fall in the bond strengths, and a reduced splitting between bonding and antibonding orbitals. It is also necessary however to take account of the considerable widths of the bands. The X-ray spectra presented in Chapter 2 (see Fig. 2.7, p. 33) show, for example, that the valence band in silicon is 15 eV broad. In the localized bonding model, the widths of the bands arise because electrons in different bonds around the same atom come quite close to each other, and so interact to produce combinations of different energy.

A slightly different view of the formation of bands in a tetrahedral solid is shown in Fig. 3.5. The horizontal axis gives the strength of the bonding interaction between neighbouring atoms. At the left, the two levels are the s and p atomic orbitals of the isolated atoms. As the orbitals overlap, bands are formed from each atomic level, with bonding combinations towards the bottom, and antibonding combinations towards the top of a band. If the bonding interaction is weak, there is a lower s band which can hold two

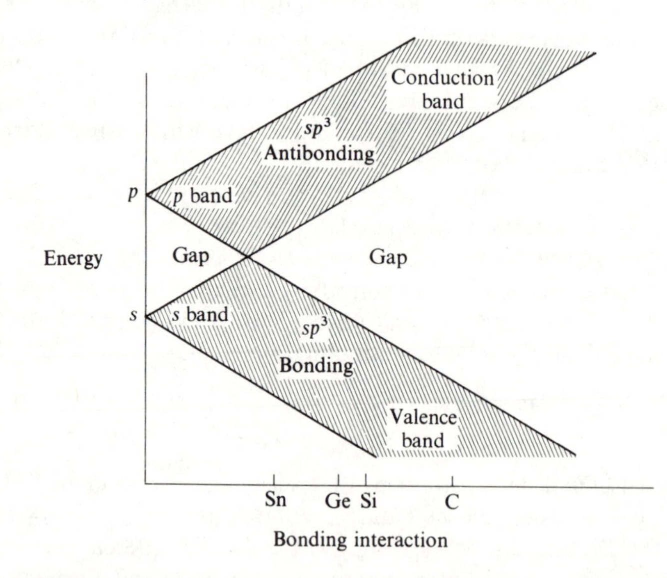

Fig. 3.5 Energy bands for a tetrahedral solid, plotted as a function of the strength of bonding between atoms. The approximate relative values of C, Si, Ge, and Sn are marked at the bottom.

electrons per atom, and the upper p band which could hold six. At a certain critical interaction the two bands cross, and this point can be regarded as the interaction strength necessary to involve s and p orbitals equally in the bonding. (We recall that the ground states of the atoms have the s^2p^2 electron configuration. In the hybrid orbital picture, it is necessary to promote one electron from the s to the p orbital. Tetrahedral bonding is favourable in these elements because the extra bond strength obtained more than compensates for this energy input.) For interactions stronger than the critical value in Fig. 3.5, there is a gap between a lower bonding band, and an upper antibonding one. In the tetrahedral diamond structure, each of these bands can now hold four electrons per atom, so that the ground state of the Group IV solids will have the lower one full, and the upper one empty. Although the valence s orbital and each of the p orbitals is involved equally in bonding in the hybrid picture, the X-ray spectrum of silicon shows that s and p orbital character is not distributed uniformly throughout the valence band. It can be seen from Fig. 2.7 on p. 33 that s character is concentrated more towards the bottom of the band, and the p orbitals contribute more to the upper part. This is to be expected from the relative energy of these atomic orbitals.

The approximate bonding strengths in the elements C, Si, Ge, and Sn are plotted in Fig. 3.5, and it can be seen that the crucial factor that determines the band gap in these elements is the strength of bonding relative to the s–p energy separation. This separation does not remain constant, but it changes much less than the bonding interaction. Thus the gap diminishes, becoming nearly zero with tin.

It is interesting to consider an element such as lead, for which the bonding is weak enough to place it to the left of the cross-over point on the diagram. Now the s band is full, and contributes nothing to the bonding, because both bonding and antibonding combinations of s orbitals are occupied. There are two electrons in the p band, and even in the unknown tetrahedral form, the element would be metallic. Since only these two electrons are effectively bonding, the metal should be thought of as divalent. In fact, there is now no advantage in adopting the tetrahedral structure, because more bonding can be obtained with a close-packed structure in which each atom has more than four near neighbours. Tin, which is very close to the cross-over point in Fig. 3.5, also has a metallic form. It appears that the reduction in the band gap down the group, and the formation of metallic structures by the lower elements, are related to the increasing tendency towards divalent chemistry, which is shown especially in tin and lead. This so-called 'inert-pair' effect has a very similar explanation: that the bond strengths in compounds become insufficient to involve the s electrons in bonding.

For elements in the later groups, it is necessary to accommodate more electrons. The tetrahedral structure is no longer the favourable one, and the compounds formed by these elements show a reduction in the coordination number. For Group V elements such as phosphorus, the pyramidal three-coordinate geometry can be explained by the existence of a pair of non-bonding electrons. Exactly the same behaviour is observed in the elemental solids. The detailed electronic structure of these elements has been investigated less extensively than the tetrahedral solids, but the principal features are undoubtedly similar to those found in molecules. The conduction bands are composed, as in Group IV, by antibonding orbitals. The top of the valence band, however, is now mainly non-bonding in character, and made up of the lone-pair orbitals on the atoms. Progressing down a Group, the band gap diminishes and a more metallic character is observed. This may be partly for the same reasons as in Group IV. However, the structures in the later Groups show more flexibility than is possible in the tetrahedral elements. For elements lower in a Group, the difference in distances between near-neighbour and next-near-neighbour atoms decreases. The greater number of overlapping orbitals will cause a broadening of the bands, and a decrease in the band gap. Thus the change over to metallic, close-packed structures happens in a more gradual way.

3.2.2 Heteropolar bonding

Very many binary solids have bonding which is intermediate between the simple ionic and covalent extremes. This class includes compounds, such as gallium arsenide, where the difference of electronegativity of the elements is quite small, as well as ones such as halides of post-transition elements where a high degree of ionic character is combined with an element of covalency. It is interesting to think of series such as CuBr, ZnSe, GaAs, and Ge, where the solids all have the same tetrahedral structure, and the same number of valence electrons. The difference of electronegativity between the atoms decreases progressively along the series, so that the bonding, which in CuBr has a strong ionic component, acquires progressively more covalent character.

It is convenient to introduce this type of solid by referring to a simple diatomic molecule AB in which there is some difference in the atomic orbital energies of A and B (Fig. 3.6). Let E_A and E_B be the atomic orbital energies of the isolated A and B atoms, and V_{AB} the interaction energy between them, which results from their overlap in the molecule. The simplest quantitative MO treatment shows that bonding and antibonding combinations formed from these orbitals will have energies given by the roots of the secular equation:

$$\begin{vmatrix} E_A - E & V_{AB} \\ V_{AB} & E_B - E \end{vmatrix} = 0. \qquad (3.5)$$

This is a quadratic equation with solutions:

$$E = (E_A + E_B)/2 \pm \{ V_{AB}^2 + (E_A - E_B)^2/4 \}^{1/2}. \qquad (3.6)$$

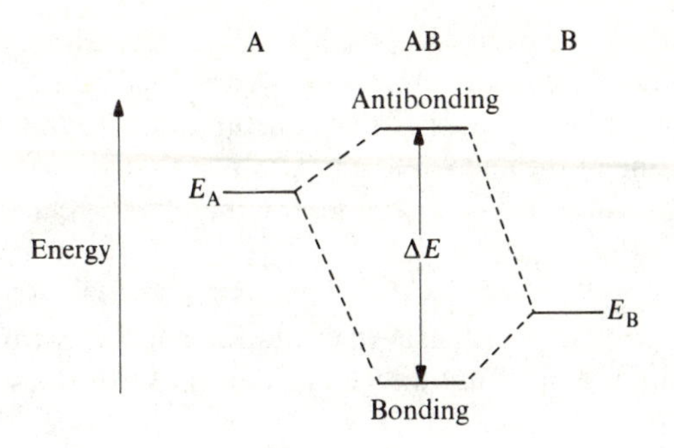

Fig. 3.6 Formation of bonding and antibonding orbitals in a heteronuclear diatomic molecule.

The energy difference between the upper antibonding and the lower bonding orbital is therefore:

$$\Delta E = \{4V_{AB}^2 + (E_A - E_B)^2\}^{1/2}. \tag{3.7}$$

This can be written:

$$\Delta E = (E_i^2 + E_c^2)^{1/2} \tag{3.8}$$

where

$$E_i = E_A - E_B \tag{3.9}$$

is the ionic contribution to the gap, being the energy difference of the isolated atoms, and

$$E_c = 2V_{AB} \tag{3.10}$$

represents the covalent part of the splitting, and is the magnitude of the gap in a homonuclear case where E_i is zero.

In a solid with covalent bonding, the bonding and antibonding orbitals form broad bands, and the gap is measured between the top of the valence band and the bottom of the conduction band. A simple equation such as 3.8 cannot unfortunately be used for the band gap itself. This equation should, however, be appropriate to the *average* energy difference between the middle of the two bands. Phillips and van Vechten showed how peaks in the absorption spectra of binary compounds could be used to determine this average excitation energy. The results of such measurements have been used to estimate E_i and E_c parameters experimentally for a range of compounds. The first step is to determine E_c for the Group IV elemental solids, where there is no ionic contribution. The values obtained (in eV) are:

C	Si	Ge	Sn
14.0	6.0	5.6	4.3

In an isoelectronic series of compounds such as Ge, GaAs, ZnSe and CuBr, the bond length hardly changes, and it can be assumed that E_c, which depends on the overlap of atomic orbitals, is also constant. For compounds that are not isoelectronic to one of the elements, it is assumed that E_c is a smooth function of interatomic distance *d*. For example, for the elements it is found that:

$$E_c \propto d^{-2.5}. \tag{3.11}$$

With values of E_c and of the measured excitation energy, E_i can be determined for each compound from equation 3.8. Values in eV for the series mentioned are:

Ge	GaAs	ZnSe	CuBr
0.0	1.9	3.8	5.6

This shows nicely how E_i, the ionic contribution to the orbital splitting, increases with the electronegativity difference between the two elements. In fact, Phillips and van Vechten were able to derive an electronegativity scale, by assuming that for a compound AB:

$$E_i = X_B - X_A \tag{3.12}$$

where X_A and X_B were characteristic parameters for the elements A and B. They should not be related to individual atomic orbital energies, but rather to average energies of the *s* and *p* orbitals used in bonding. Selected values of Phillips–van Vechten electronegativities X are shown in Table 3.1, together with values from the Pauling scale. There is a fair correlation between the values for non-metallic elements, but not such good agreement for the metals. In fact the Pauling electronegativity scale was derived from a formula involving covalent bond strengths. This formula does not work well for compounds of metallic elements, and Pauling's values for these elements are probably not very meaningful.

As well as predicting the energy gap between bonding and antibonding molecular orbitals, the simple molecular orbital model gives the relative atomic orbital contributions to each molecular orbital. The bonding orbital has a larger contribution from the more electronegative atom, that is B in Fig. 3.6. The squares of the atomic orbital coefficients for this molecular orbital provide a definition of the fractional ionic character in the bond, and in terms of the parameters used earlier, this comes out as:

$$f_i = E_i/\Delta E = E_i/\{E_i^2 + E_c^2\}^{1/2}. \tag{3.13}$$

Fractional ionic characters f_i for some halides, using E_i and E_c parameters determined from spectroscopic measurements as described, are found to be:

LiCl	NaCl	KCl	CuCl	AgCl
0.90	0.94	0.95	0.77	0.86

Table 3.1
Comparison of Phillips–van Vechten and Pauling electronegativities

Element	*Phillips—van Vechten*	*Pauling*
Na	0.7	0.9
Cu	0.8	1.9
Ag	0.6	1.9
Mg	1.0	1.2
Zn	0.9	1.6
Cd	0.8	1.7
O	3.0	3.0
S	1.9	2.5
Se	1.8	2.4
F	4.0	4.0
Cl	2.1	3.1
Br	2.0	2.8

Values for the alkali halides are satisfyingly high, those for the copper and silver compounds being rather lower, as we might expect.

One of the most remarkable features of the Phillips–van Vechten model is the correlation found between the structure type and the fractional ionic character. In the simple ionic model, the crystal structure is supposed to be determined by the relative sizes of the anion and cation—the so-called radius ratio. Even for alkali halides, however, the radius ratio rules do not work perfectly in a quantitative way. They become even less reliable when applied to compounds with a lower degree of ionic character. Many AB compounds formed by post-transition metals adopt the tetrahedral zinc-blende or wurtzite structures, rather than the sodium chloride structure predicted by the ionic model. The reason must be that covalent bonding with *s* and *p* valence orbitals favours tetrahedral coordination, as in the Group IV elements. Figure 3.7 shows a plot of the E_i and E_c energy parameters for a large number of AB compounds. A straight line from the origin on this plot corresponds to a constant ratio E_i/E_c, and hence from equation 3.13, to a constant fractional ionic character f_i. It is remarkable that the compounds with tetrahedral structures are separated almost perfectly from those with the NaCl structure. The critical value of f_i is apparently 0.785, and compounds with f_i lower than this have the zinc-blende or wurtzite structure. It can be seen that the wurtzite structure occupies an intermediate position, being adopted by slightly more

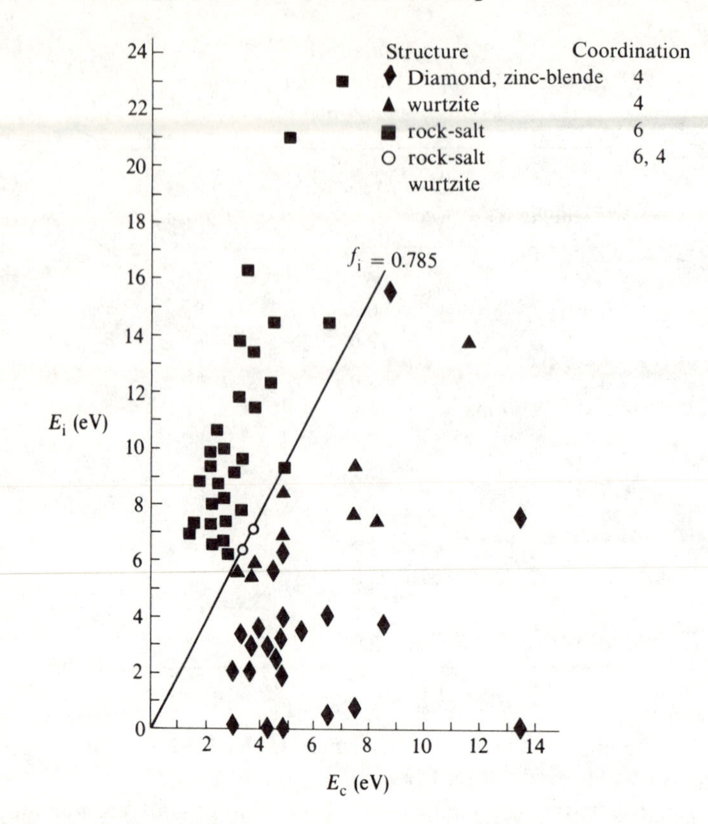

Fig. 3.7 Phillips–van Vechten plot for binary compounds AB. The E_i and E_c parameters are defined in the text, and the straight line corresponds to an ionicity of 0.785. (After J. C. Phillips, *Revs. Mod. Phys.*, **42** (1970), 317.)

ionic compounds than those with the zinc blende structure. The difference between the two structures lies in the relative positions of more distant atoms. As a consequence, wurtzite has a slightly larger Madelung constant than zinc blende, and is favoured by more ionic compounds.

3.2.3 Dielectric properties of binary compounds

We saw earlier in Section 2.4.2 (p. 39), that the way in which the dielectric 'constant' varies with frequency has an important bearing on the optical properties of a solid. Dielectric properties are also important in other respects, as they determine the screening of electrostatic forces between extra charges in the lattice. We shall see in Chapter 7 that the behaviour of charged defects, and of electrons and holes introduced by impurities or by optical excitation, all depend on the dielectric constants. It is useful at this point to explore more fully how the dielectric properties of a non-metallic solid depend on the kind of bonding.

A sketch of how the dielectric function changes in the region of an electronic absorption band was given in Fig. 2.12, on p. 40. In addition to electronic excitations, solid compounds also have infra-red active vibrational modes. Known as **optical phonons**, these modes correspond to the out-of-phase motion of the different atoms, producing an oscillating dipole as with infra-red active molecular vibrations. The dielectric function changes in the same way near an infra-red active vibrational frequency as it does near an electronic absorption band. A schematic picture of the dielectric function, including both vibrational and electronic excitations, is given in Fig. 3.8. There are two important frequency regions for practical purposes:

1. The high frequency, or optical, dielectric constant, ε_{opt}, is measured in a region below any electronic absorption, but above the vibrational frequency. At these frequencies, only the electrons can respond effectively to an oscillating electric field, so that ε_{opt} is determined solely by the electronic polarizability of the solid.
2. The static dielectric constant, ε_s is taken at frequencies well below the vibrational absorptions. The polarization of the solid in an electric field at these low frequencies has an additional contribution from the motion of ions as a whole, so that ε_s is normally larger than ε_{opt}.

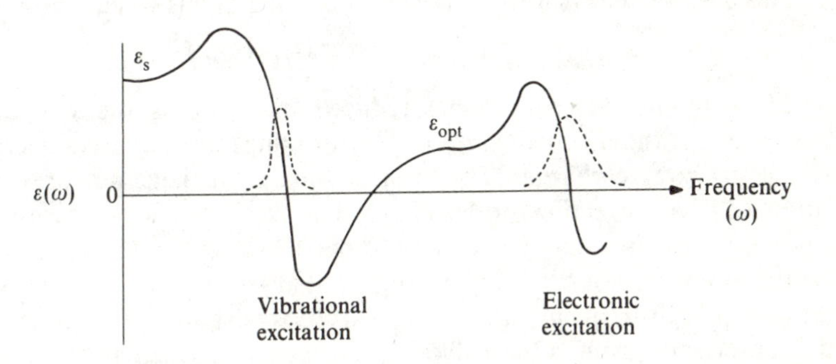

Fig. 3.8 Real (ε') and imaginary (ε'') parts of the dielectric function of a non-metallic solid, showing electronic and vibrational absorption regions. ε', ——; ε'', - - - - -.

It is ε_{opt} that determines the refractive index at frequencies below the electronic absorption edge; indeed measurement of the refractive index is the best way of estimating it. Adapting equation 2.6 of Chapter 2 (p. 39) gives an approximation to ε_{opt} well below the optical absorption frequency:

$$\varepsilon_{opt} - 1 = Ne^2/(m\varepsilon_0\omega_e^2) \tag{3.14}$$

where ω_e represents an average electronic excitation energy. Although there is no reason why such an average energy should correlate perfectly with the band gap, there is a tendency for solids with smaller gaps to have an higher value of ε_{opt}. This is illustrated in Table 3.2, where the band gaps and dielectric properties of some simple solids are given.

Table 3.2

Dielectric properties of some simple solids

Solid	*Band gap,* E_g (eV)	*Dielectric constants* ε_{opt}	ε_s	*Transverse charge,* e_T
NaCl	8.5	2.3	5.6	1.1
CuBr	3.5	4.0	7.0	1.5
ZnSe	2.8	5.4	9.2	2.0
GaAs	1.5	10.6	11.3	2.2
Ge	0.7	16.0	16.0	0

The additional contribution to the static dielectric constant, which comes from the displacement of atoms in the solid under the influence of an electric field, can be estimated by a formula somewhat analogous to equation 3.14:

$$\varepsilon_s - \varepsilon_{opt} = [(\varepsilon_{opt} + 2)/3]^2 N e_T^2/(M \varepsilon_0 \omega_v^2) \tag{3.15}$$

Now M represents the vibrational reduced mass, ω_v is the vibrational frequency, and e_T the effective charge on the vibrating ions, which is sometimes called the **transverse charge**. (The term in square brackets, not present in equation 3.14, is a correction for the influence of the electronic polarization on that produced by the ions.) The data in Table 3.2 show that for germanium, where the atoms do not carry any charges so that there is no infra-red active vibration, the high-frequency and static dielectric constants are the same. For all the compounds given in the Table 3.2, ε_s is higher than ε_{opt} as expected. In the case of a highly ionic compounds such as NaCl, the transverse charge deduced from the dielectric data is close to one. However, other compounds seem to have unrealistically high values. It must be remembered that when covalent bonding is present, vibrations do not involve the motion of fixed charges, but rather the stretching of covalent bonds. What is being measured is in fact the **dynamic dipole moment**, which determines the strength of the infra-red absorption. In molecular vibrations the dynamic dipole does not correlate well with the static charge distribution, but rather reflects the way in which electrons are redistributed during the bond stretching. The same is true in solids, and although the transverse charge is a useful parameter for interpreting the dielectric properties, it cannot in general be used as a guide to the static electron distribution.

3.3 Metals

3.3.1 Simple metals—the free electron model

Metallic solids are generally formed by elements towards the left-hand side of the Periodic Table, where the ionization energies are fairly low, and where the number of electrons may be less than the number of valence orbitals available to them. Close-packed or nearly close-packed structures are formed for the same reason as the clusters shown by electron-deficient molecules such as boron hydrides: when there are relatively few valence electrons, the maximum bonding is achieved by increasing the number of near-neighbour atoms. Most non-metallic elements, on the other hand, have a greater proportion of valence electrons to orbitals. In this electron-rich situation, close-packed structures are unfavourable, as some electrons would have to be accommodated in antibonding orbitals. As we have seen, non-metallic elements generally adopt structures of lower coordination, where electrons can occupy bonding and non-bonding orbitals, and antibonding levels are left empty.

A simple picture of the energy levels in a metal such as sodium would suggest that separate bands are formed from the 3*s* and 3*p* valence orbitals. If this were true however, in the divalent element magnesium the 3*s* band would be full, and the solid would not be metallic. In fact the valence orbitals of the atoms early in the period are very diffuse, and overlap strongly with neighbouring atoms. The bands produced are many electron volts broad—very wide compared with the energy separation of different atomic orbitals. Thus the bands formed from atomic 3*s*, 3*p*, and even higher orbitals overlap in energy, and merge into a single wide conduction band. In a real sense, the atomic orbitals lose their identity, and the electrons in the metal hardly feel the potential of individual atoms. The most successful simple model of these metals is based, not on the chemical picture of overlapping atomic orbitals, but on the idea of free electrons moving in a nearly constant potential. The free electron model works best for the pre-transition metals of the A sub-Groups I, II, and III. For post-transition elements in the B sub-Groups, the interaction between atomic orbitals in the solid is not quite so large, and the free electron theory does not work so well. The most serious breakdown of the model however, occurs with the transition metals, which are discussed separately in the next section.

The 'free' electrons in a simple metal behave rather like a gas, but a rather unusual one with properties dominated by the exclusion principle. In a conventional gas of molecules at normal temperature and pressure, the number of molecules is vastly exceeded by the number of translational energy levels available to them. In the calculation given below, we shall see that the number of translational levels in a given energy range depends on the mass of the particles in the gas, and on their density. Electrons are much lighter than molecules, and the density of conduction electrons in a metal is also much higher than that of a conventional gas. It is these two differences that lead to the unusual properties of the electron gas.

Suppose that the electrons are confined to a cube of metal, with sides of length a. Within the cube, the potential is assumed to be constant, and may be taken as zero. Thus the energy levels are simply those of a particle in a box:

$$E = (n_x^2 + n_y^2 + n_z^2)h^2/(8ma^2). \tag{3.16}$$

n_x, n_y, and n_z can each take the values 1, 2, 3, . . . , and represent the number of half waves which fit into the x, y, and z directions in the cube. It is convenient to represent equation 3.16 in a graphical way by thinking of each combination of quantum numbers (n_x, n_y, n_z) as a point in a cubic lattice. A two-dimensional representation of this is shown in Fig. 3.9. Each point in the lattice represents one possible set of quantum numbers, and therefore corresponds to one orbital state in equation 3.16. According to the exclusion principle, each such orbital can hold two electrons with opposed spins. If the spacing of points in the imaginary lattice is one unit, the point representing a particular state is found at a distance R from the origin given by:

$$R^2 = n_x^2 + n_y^2 + n_z^2 = 8mEa^2/h^2 \tag{3.17}$$

Thus all states with a given energy E are represented by points in the lattice at the same distance R from the origin. Since only positive values of n_x, n_y, and n_z are allowed, the number of orbitals below a certain energy E_{max} can be found by counting the points inside the positive octant of a sphere of radius R_{max},

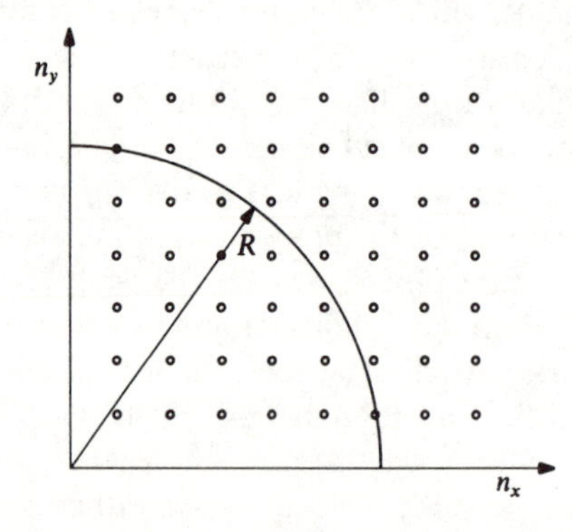

Fig. 3.9 Two-dimensional representation of how each free electron state can be represented as a point (n_x, n_y) on a lattice. The distance R from the origin gives the energy of a state according to equation 3.17.

given in terms of E_{max} by equation 3.17. Since each point is at the corner of a unit cube in the imaginary lattice, the number of points is simply the volume of the octant. The number of electrons N that can be accommodated in energy levels up to E_{max} is twice this:

$$\begin{aligned} N &= 2(1/8)(4/3)\pi R_{max}^3 \\ &= 8\pi/3(2mE_{max}/h^2)^{3/2}a^3. \end{aligned} \tag{3.18}$$

The factor a^3 is the volume of the box in which the electrons are contained, and so E_{max} can be written in terms of the density of electrons ρ, equal to N/a^3:

$$E_{max} = h^2/(2m)(3\rho/8\pi)^{2/3}. \tag{3.19}$$

This is the energy of the highest occupied level at absolute zero, with all orbitals doubly filled according to the exclusion principle. For the conduction electrons in metallic sodium, $\rho = 2.5 \times 10^{22}$ cm^{-3}, and the predicted value of E_{max} is 3.2 eV. At room temperatures kT is 0.024 eV, and so only a small minority of electrons close to top-filled level can be thermally excited. By contrast, for gaseous nitrogen at its normal density, E_{max} is around 10^{-6} eV, so that virtually all molecules will be in thermally excited translational levels. It is only at extremely high densities (for example, those found near the middle of some stars) that the calculation of E_{max} is relevant for a gas of atoms or molecules.

It is interesting to compare the values of E_{max} calculated from the free electron theory with measurements (from photoelectron or X-ray absorption spectroscopy) of the width of the occupied part of the conduction band. The results for some simple metals are:

E_{max} (eV)	Na	Mg	Al
Calculated	3.2	7.2	12.8
Experimental	2.8	7.6	11.8

The agreement is very good, and indeed close to the limits of accuracy of these measurements.

The free electron model also allows the density of states $N(E)$ within the conduction band to be calculated. $N(E)$ is the number of states available for electrons per unit volume per unit energy range, and therefore the total electron density is the integral of $N(E)$ up to the top-filled level E_{max}. By differentiating equation 3.18, which gives the total number of electrons, we find:

$$N(E) = 4\pi(2m/h^2)^{3/2}E^{1/2}. \tag{3.20}$$

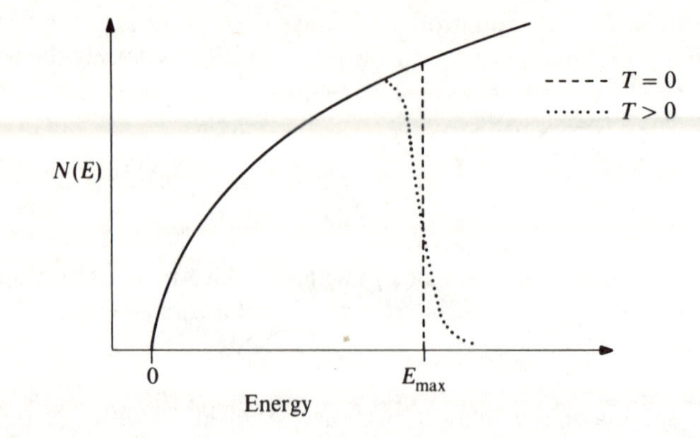

Fig. 3.10 Density of states in the free electron model. The Fermi–Dirac functions showing the occupancies of levels at absolute zero and at some higher temperature are indicated.

This function is shown in Fig. 3.10. The photoelectron spectra of many simple metals (for example that of aluminium in Fig. 2.4 on p. 30) give at least an approximate confirmation of this prediction.

Figure 3.10 also illustrates how the levels are occupied by electrons at absolute zero, and at some higher temperature. As was mentioned in Chapter 1 (see equation 1.11), at finite temperatures the fractional occupancy of levels is given by the Fermi–Dirac distribution function:

$$f(E) = \{1 + \exp[(E - E_F)/kT]\}^{-1} \tag{3.21}$$

It is only within an energy range of the order of kT from the Fermi level E_F that electrons can be thermally excited, and for most metals at temperatures below their melting point, this only covers a small fraction of the occupied levels. For this reason the thermodynamic properties of metallic electrons are very different from those of a normal gas. For example, in an ideal monatomic gas, where each atom makes an independent contribution to the total translational energy, the specific heat at constant volume C_v, has the value $3R/2$. In the electron gas, only electrons close to the Fermi energy can be excited thermally, and the specific heat is much lower. Like many other properties, C_v depends not on the number of electrons present, but on the density of states at the Fermi level, $N(E_F)$. A calculation based on the Fermi–Dirac distribution gives:

$$C_v = \pi^2/3N(E_F)k^2T \tag{3.22}$$

The specific heat of a solid also has contributions from lattice vibrations, which normally swamp the electronic term. But at low temperatures the lattice term is proportional to T^3, and so as the temperature is lowered, it disappears faster than the electronic one. Figure 3.11 shows a plot of C_v/T against T^2 for

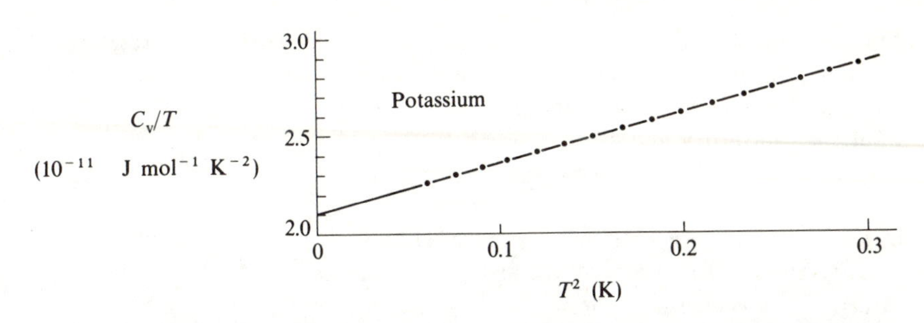

Fig. 3.11 Specific heat for potassium at low temperatures, plotted as C_v/T against T^2. The intercept at $T = 0$ gives the electronic contribution. (After W. H. Lien and N. E. Phillips, *Phys. Rev.*, **133** (1964), A1370.)

potassium, measured at temperatures below 1 K. The slope comes from the T^3 lattice term, and the intercept at $T = 0$ gives the electronic contribution. Measurement of the electronic specific heat is one of the best ways of determining the density of states at the Fermi level.

Another property that depends on electrons close to the Fermi level is the paramagnetic susceptibility of metallic electrons. The spin-up and spin-down states of an electron have different energies when a magnetic field is applied, and the effect of this on the electrons in a metal is illustrated in Fig. 3.12. The density of states is shown separated into parts for spin-up and spin-down electrons. In the absence of a magnetic field, the two curves are identical.

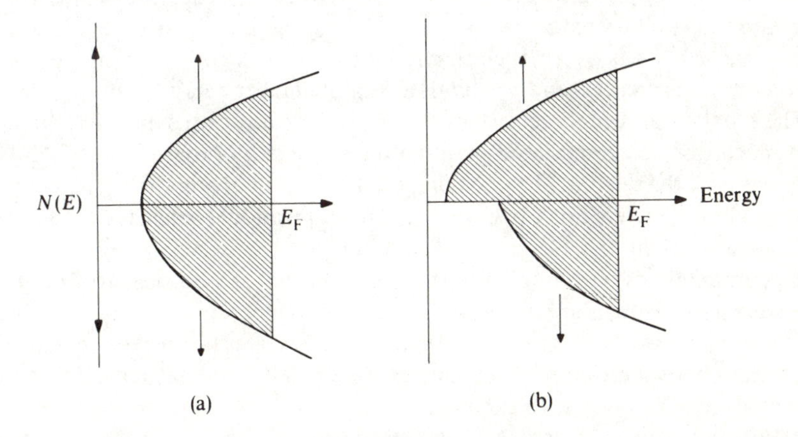

Fig. 3.12 Effect of applying a magnetic field to a simple metal. Separate density-of-states curves are shown for spin-up (↑) and spin-down (↓) electrons. (a) Spin-up and spin-down electrons have the same energies with no field; (b) the levels are shifted in a magnetic field, causing an excess of spin-up electrons. The shift is greatly exaggerated.

Application of a magnetic field gives a shift in the energies (greatly exaggerated in the diagram). The ground state of the metal now has excess of electrons of one spin. The effect is proportional to $N(E_F)$, and a calculation of the so-called **Pauli susceptibility** gives:

$$\chi = 2\mu_0\mu_B^2 N(E_F), \tag{3.23}$$

where μ_B is the Bohr magneton. Figure 3.12 shows the situation at absolute zero. Increasing the temperature has only a very small effect, and the Pauli susceptibility is essentially temperature-independent. This contrasts with the $1/T$ Curie law found for localized unpaired electrons. The Curie law arises when the lining up of electron spins in a magnetic field is opposed by thermal agitation, but this effect is not important in the electron gas.

The metallic elements for which the free electron theory works best are often described as 'simple metals'. These form a relatively small class comprising the alkali and alkaline earth metals, and aluminium. Because of its physical simplicity, however, the theory is often used as a framework for interpreting the properties of many other solids. The justification for applying the free electron model more widely than it seems to deserve, and the corrections which sometimes need to be made to it, are discussed again in Chapter 4.

3.3.2 Transition metals

Among metallic elements, it is in the transition series that the free electron theory breaks down most seriously. The *d* valence orbitals, especially the 3*d* in the first series, are more contracted than valence *s* and *p* orbitals, and do not overlap so strongly. The result is a narrower band, which is more identifiable in its atomic orbital character than with the simple metals. Figure 3.13 shows how the *d* band is overlapped in energy by a broad, free-electron-like band, arising from valence *s* and *p* orbitals at slightly higher energy than the *d* levels. The *d* band has a high density of states, because altogether ten electrons per atom can be accommodated within a narrow energy range. Figure 3.13 also shows how the *d* band fills across a transition series, giving the position of the Fermi level for elements early (such as Ti), in the middle (such as Mn) and late (such as Ni) in the 3*d* series. The filling of the band can be observed experimentally, for example by the combination of photoelectron and inverse photoelectron spectra shown previously in Figure 2.6 (p. 32). These spectra also show that the *d* band becomes narrower, as the increase of effective nuclear charge across the series causes the *d* orbitals to contract, so that they overlap less strongly with each other.

Within a band such as that formed by *d* orbitals, states near the bottom are bonding between adjacent atoms, and states near the top antibonding. We would expect the maximum bonding from the *d* electrons to be found near the middle of the series, when the *d* band is half-full. Figure 3.14 shows the sublimation energies for the elements of the three transition series. For the

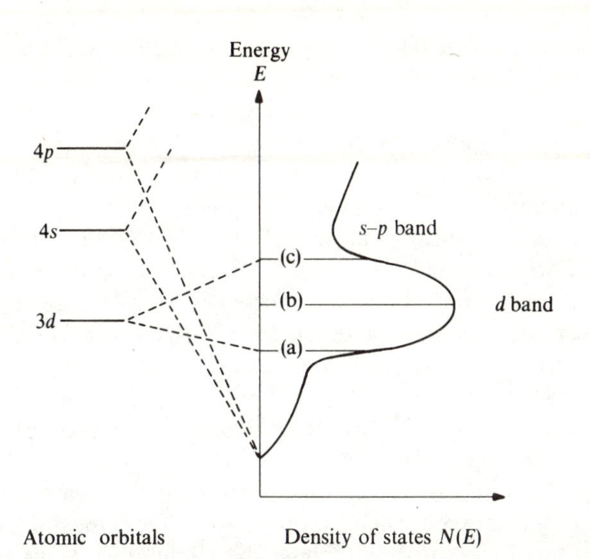

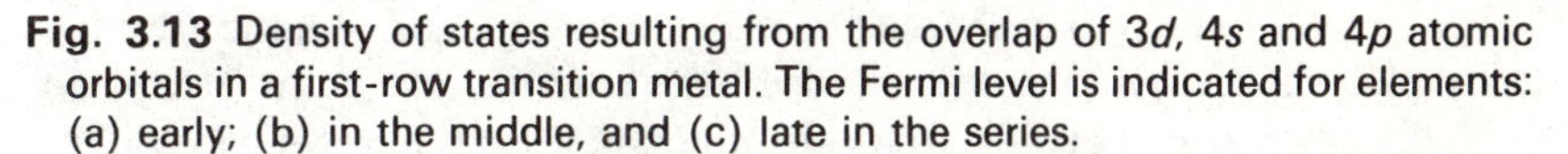

Fig. 3.13 Density of states resulting from the overlap of $3d$, $4s$ and $4p$ atomic orbitals in a first-row transition metal. The Fermi level is indicated for elements: (a) early; (b) in the middle, and (c) late in the series.

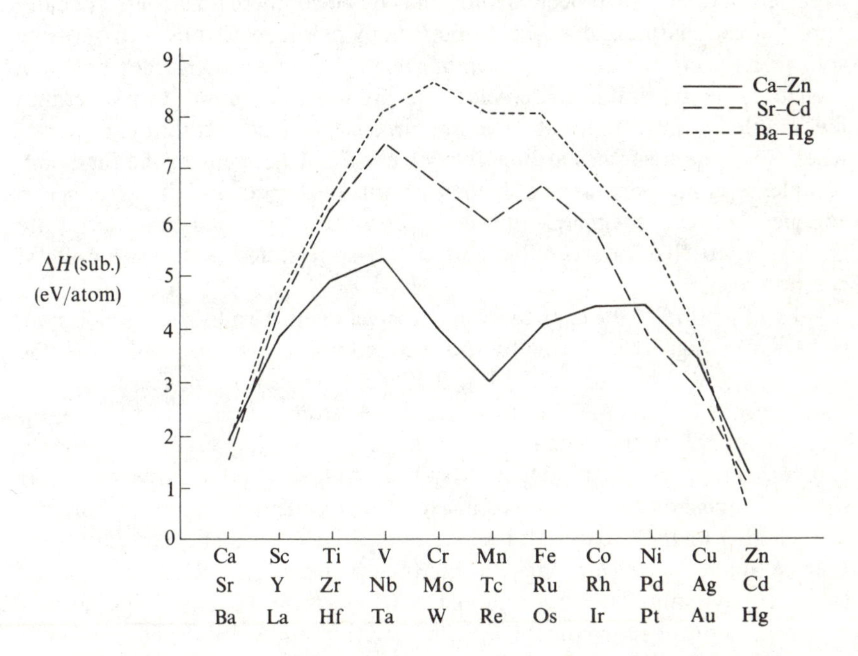

Fig. 3.14 Sublimation energies for transition metals of the three series.

second and third series there is a peak near the middle. The sublimation energies in the $3d$ series are lower than the other two, reflecting the relatively poor overlap obtained between $3d$ orbitals, and the weaker bonding which results. This series also has a pronounced dip in the middle. The elements around iron seem to have lower bonding energies than would be expected. There is an interesting connection between this and the magnetic properties of the metals. It is well known that iron, cobalt, and nickel are ferromagnetic below their Curie temperatures. The earlier metals, chromium and manganese, have more subtle antiferromagnetic properties. In each case the magnetism arises because atoms in the solid do not have the same number of spin-up and spin-down electrons, and so have a magnetic moment. In the antiferromagnetic case, moments on different atoms point in different directions so as to give a net cancellation, but in the ferromagnetic metals, they are aligned parallel to one another.

The magnetic moments of the transition-metal atoms are a result of the strong repulsion between electrons in d orbitals on the same atom. This electrostatic repulsion can be decreased by aligning electrons with parallel spins, so that they avoid each other spatially in accordance with the exclusion principle. The energy lowering is called the **exchange energy**. In free atoms, this effect gives rise to **Hund's first rule**, which states that the ground state of a given electron configuration is obtained when as many electrons as possible have parallel spins. In molecules and solids however, there is an energy penalty to pay, since maximum bonding is obtained by pairing electrons with opposite spin in bonding orbitals. The bonding energy is normally larger than the exchange energy, so that paired electrons are more common. It is sometimes favourable for an atom to retain an unpaired electron configuration, however, when the bonding forces are relatively weak. This happens in the high-spin complexes, commonly formed by the elements of the $3d$ series. The balance in energies between magnetic and non-magnetic states of a metal is quite analogous to that between high- and low-spin states in transition-metal compounds.

A band picture of the change from a non-magnetic to a ferromagnetic state is shown in Fig. 3.15. Normally, spin-up and spin-down electrons have the same energy within a band (see Fig. 3.15(a)). However, in the ferromagnetic state in Fig. 3.15(b), some electrons have shifted from the spin-down to spin-up levels. Because of the excess of spin-up electrons, they each experience a smaller average repulsion, so that the spin-up states are stabilized as shown. At the same time, however, it is necessary to move electrons to more antibonding levels higher up in the band, and some bonding energy is lost. Since the d band holds a total of ten electrons per atom, the transfer of one electron, from the top occupied spin-down level to the bottom empty spin-up level, costs a bonding energy of approximately one-fifth of the bandwidth W. The repulsion energy is reduced by an amount equal to the exchange integral K between

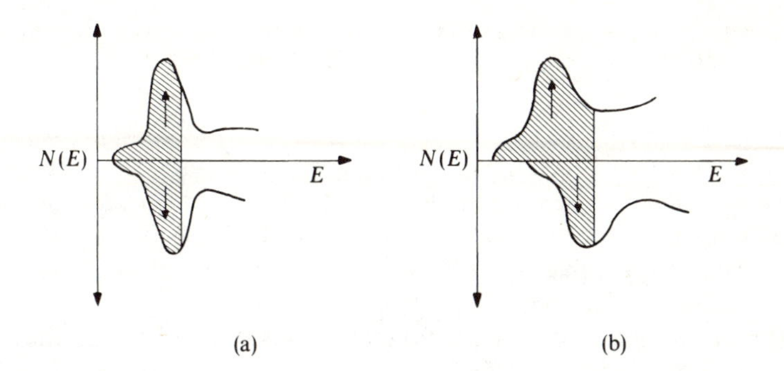

Fig. 3.15 Band picture of a ferromagnetic transition metal: (a) non-magnetic state with equal numbers of spin-up and spin-down electrons; (b) ferromagnetic state.

electrons in *d* orbitals on the same atom. The ferromagnetic state will only be stable if there is a net energy lowering, that is if the reduced repulsion compensates the loss in bonding energy. This requires:

$$K > W/5. \tag{3.24}$$

The smallest bandwidths are found for the later elements of the 3*d* series, where the orbitals are contracted and the overlap is weak. The same contraction also brings electrons on one atom closer together, thus increasing the electron repulsions and hence the exchange integral K. It can be understood therefore, why the unusual magnetic properties, and the reduced sublimation energies which go with these, are shown by the later metals of the first transition series. For elements of the second and third transition series, the more extended nature of the 4*d* and 5*d* orbitals gives larger bandwidths and smaller exchange integrals. Thus similar magnetic properties are not found. Ferro- and antiferro-magnetism is shown, however, by some metallic transition-metal compounds, when the *d* band (discussed in the next section) is fairly narrow. For example, chromium dioxide, CrO_2, is a ferromagnetic metal, used widely in magnetic recording tapes.

3.4 Transition-metal compounds

With their partially filled *d* levels, transition-metal compounds show a much more varied range of electronic behaviour than the simple solids considered so far in this chapter. Some of these properties require a more detailed treatment, and are considered further in later chapters. However, it is not difficult to describe the principal features of the *d* levels in solids, using the same chemical arguments as in the preceeding sections. The compounds discussed here are those where the transition metal is combined with an element of fairly high electronegativity, such as oxygen. The best starting point is the ionic model of Section 3.1.1.

3.4.1 The *d* band

The basic energy level diagram for a transition-metal oxide is similar to that for other ionic compounds, with a valence band of oxygen $2p$ character, and a conduction band of metal character. However, the lowest metal orbitals are normally the *d*, rather than *s* as in the pre-transition elements. In Fig. 3.16, the *d* orbitals are shown forming a single band, separated from those of the higher energy *s* and *p* orbitals of the transition metal. As we shall see later, the *d* band may be split by ligand field effects, similar to those found in transition metal complexes.

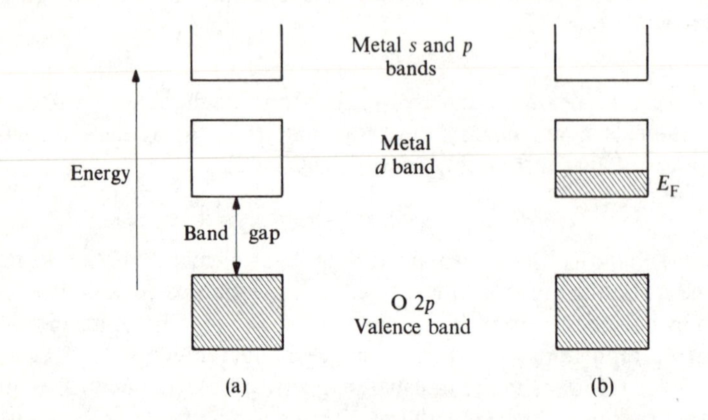

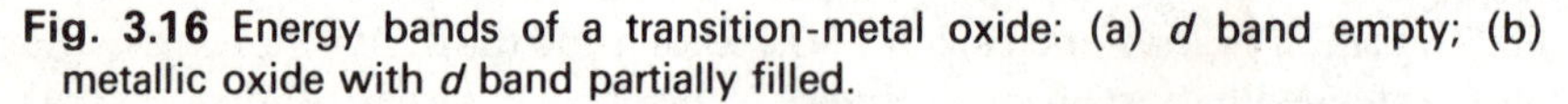

Fig. 3.16 Energy bands of a transition-metal oxide: (a) *d* band empty; (b) metallic oxide with *d* band partially filled.

In the ionic compounds considered earlier in this chapter, the conduction band is normally empty giving an insulator, but in transition-metal compounds there may be electrons present in the *d* conduction band. The occupancy of the *d* band can be found from the same chemical arguments that are used for *d* electron configurations in complexes. Thus Ti^{4+} has the $3d^0$ configuration, and so TiO_2 has no electrons in the *d* band. The observed gap, 3 eV, is fairly small, and suggests the presence of an appreciable covalent mixing between the atomic orbitals of oxygen and the metal. As with molecules and complexes, it would be more correct to describe the valence band as a bonding combination of oxygen and metal orbitals, and the 'metal *d*' conduction band as an antibonding combination. However, the description in terms of an oxygen and a metal band is useful, as it denotes the principal atomic orbital constituents of each level. Whereas TiO_2 is an insulator, the compounds Ti_2O_3 and VO_2 have one electron per atom in the *d* band (see Fig. 3.16(b)), and are metallic in character, at least at higher temperatures. Thus in simple cases, the electronic nature of a transition-metal compound can be understood in terms of the *d* electron configuration of the metal atom. Another pair of compounds which illustrates this argument is that of WO_3 ($5d^0$ and insulating) and ReO_3 ($5d^1$ and metallic).

Ternary compounds, containing a pre-transition metal as well as a transition metal, can be treated in the same way. For example, in the sodium tungsten bronze compounds, Na_xWO_3, the sodium 3*s* orbitals are of considerably higher energy than the tungsten 5*d*. The energy level diagram of Fig. 3.16 can still be used therefore, and only the higher regions of the conduction levels will have any sodium character. The extra valence electron introduced by each sodium will go into the W 5*d* band. For compositions with *x* larger than about 0.3, the compounds are metallic. The photoelectron spectra of $Na_{0.7}WO_3$ shown in Fig. 2.5 (p. 30) confirm this picture of the electronic structure. The sodium tungsten bronze is one of a wider class of oxide bronzes, in which insertion of an alkali metal or hydrogen introduces conduction electrons into a d^0 transition-metal oxide. These compounds have an interesting range of electronic properties. With a very small number of additional electrons, they are often not metallic. As discussed in Chapter 6, it is likely that the electrons are trapped by lattice distortions, so that they are unable to conduct in a metallic fashion, and give rise instead to semi-conducting properties.

3.4.2 Ligand field splittings

It is well known in transition-metal complexes that the *d* orbitals are split in energy by the surrounding ligands. This effect is not primarily due to electrostatic repulsion, as was originally thought, but comes from the bonding interaction of *d* orbitals with the ligand atoms. The '*d* orbitals', as we suggested above, are really antibonding combinations of metal *d* with ligand orbitals. Those *d* orbitals which point directly towards ligand atoms form σ combinations, with a stronger degree of overlap than the π combinations formed by the *d* orbitals pointing between ligand atoms. The σ antibonding combinations are higher in energy, and this forms the major contribution to the ligand field splitting. The energy ordering of the *d* orbitals depends on the geometry of the ligands surrounding a transition-metal atom, and three cases which we discuss briefly below are illustrated in Fig. 3.17.

In the octahedral case shown in Fig. 3.17(a), three *d* orbitals, called the t_{2g} set, are lower in energy than the other two, the e_g set. In a solid where the transition-metal atom occupies an octahedral site, the separate levels will be broadened into bands. The bandwidth may be less than the ligand field splitting, giving a gap between the lower t_{2g} band and the upper e_g. In this case a compound with six *d* electrons could have a full t_{2g} band and be non-metallic. In chemical terms, we would have a d^6 low-spin compound, and in the first transition series this is most likely to be found with Co^{3+}. The compound $LaCoO_3$,

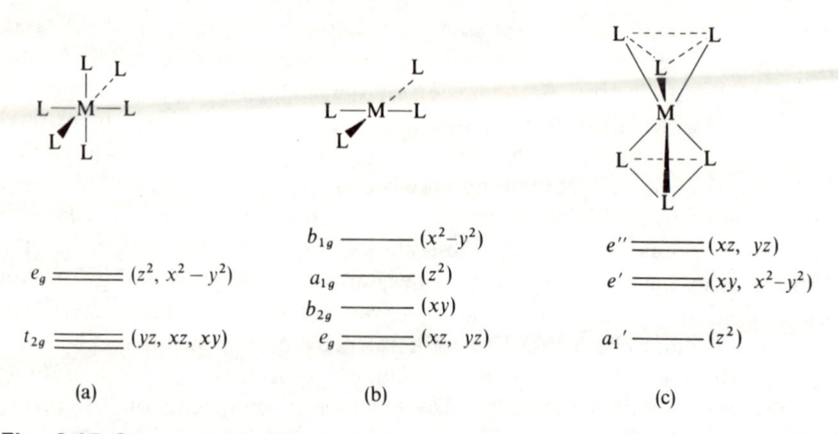

Fig. 3.17 Some possible ligand geometries, and resulting *d*-orbital splittings: (a) octahedral coordination; (b) square-planar coordination common in Pt^{2+} compounds; (c) trigonal prismatic coordination of MoS_2.

which has the perovskite structure with Co^{3+} occupying an octahedral site, is indeed non-metallic at low temperatures. However, it appears that the gap between the top of the t_{2g} and the bottom of the e_g bands is very small here, and at high temperatures $LaCoO_3$ undergoes a complicated electronic transition. As might be expected from the behaviour of transition-metal complexes, larger ligand field splittings are found in the second and third transition series, and the $4d^6$ compound $LaRhO_3$ is non-metallic with a band gap of 1.6 eV. Other d^6 octahedral compounds, such as FeO, have a high-spin configuration, in which the exchange interaction between the d electrons makes it more favourable for them to have parallel spins, at the expense of occupying higher energy orbitals. As we shall see later, the simple band picture is less useful in thinking about high-spin compounds such as this.

There are other geometries that give rise to low-spin compounds with filled bands. An important one is the square-planar arrangement of ligands (see Fig. 3.17(b)), common for low-spin d^8 compounds such as those of Pt^{2+}. Thus $K_2Pt(CN)_4 \cdot 3H_2O$ is an insulator, with quite a large gap above the top-filled band formed from the d_{z^2} orbital. The square-planar units stack in the crystal, to form chains of platinum atoms. There is quite a large overlap between d_{z^2} orbitals on neighbouring Pt atoms, and a considerable width to the band. In fact, it is possible to remove some electrons from the band by partial oxidation, as for example in the compound $K_2Pt(CN)_4Br_{0.3} \cdot 3H_2O$. Bromine enters the structure as the Br^- ion, and gives rise to 0.3 holes per Pt atom. The compound is metallic, but it is clear from its properties that the metallic electrons are free to move only along one axis in the crystal, the direction of the Pt chains. Such 'one-dimensional' metals have unusual properties, and are discussed in Chapter 6.

Another interesting example of ligand field effects is found in the compound MoS_2, which has a layer structure, where S–Mo–S 'sandwiches' are held together by van der Waals' forces. The CdI_2 layer structure was illustrated earlier, in Fig. 1.5 (p. 9). In CdI_2, which is the structure adopted by TiS_2 and some other disulphides, the metal atoms have octahedral coordination. MoS_2 is quite similar, except that the arrangement of sulphur atoms around the metal forms a trigonal prism as shown in Fig. 3.17(c). This geometry gives a ligand field splitting with a single *d* orbital lowest in energy. The d^1 compounds NbS_2 and TaS_2 are metallic, but MoS_2 is not, because it has two *d* electrons, which fill the band formed from the lowest *d* orbital.

3.4.3 Metal–metal bonding in compounds

Many transition-metal compounds, particularly those which are metallic, have *d* bands a few electron volts in width. Some of the compounds discussed above have structures in which the metal atoms are too far apart for substantial overlap to occur directly between their *d* orbitals. Thus in compounds such as ReO_3 the *d* bandwidth cannot come from such direct overlap. Instead, it is a consequence of indirect covalent interactions via an intervening oxygen atom. The way in which such indirect interactions can give rise to bands is explained in Chapter 4. The important point here is that the metallic nature of these compounds is not a result of metallic bonding in the normal sense. There are other compounds, however, where direct overlap of *d* orbitals does occur, giving rise to metal–metal bonding in the solid. This probably happens, for example, in the metallic oxides TiO and VO, which have structures based on rock-salt. Although in the sodium chloride structure the metal atoms are partly separated by oxygens, TiO and VO both contain a large proportion of lattice vacancies. These vacancies allow the lattice to contract, thus decreasing the distance between metal atoms and allowing greater overlap between them.

The clearest examples of metal–metal bonding, however, occur in the class of compounds known as 'metal rich', in which the transition metal has an oxidation state lower than that normally expected. Many of these compounds have unusual structures, showing extended metal–metal bonding. One example is ZrCl, which was illustrated in Fig. 1.5 (p. 9). The structure has a double layer of zirconium atoms sandwiched between chlorines. The best simple description of the electronic structure is to think of the chlorine bonded as Cl^- ions. The three remaining valence electrons on each zirconium are used for metallic bonding, which is provided by direct overlap between the metal *d* orbitals. These orbitals still form a band above the anion valence band, as in the simpler cases discussed above. Figure 3.18 shows the photoelectron spectrum of ZrCl, in which one can clearly see the Cl 3*p* valence band, and the occupied Zr 4*d* band formed by the metallic electrons. Many metal-rich compounds are now known, where the metallic bonding is displayed in layers or chains of metal atoms. These extended solids can be compared with cases like the

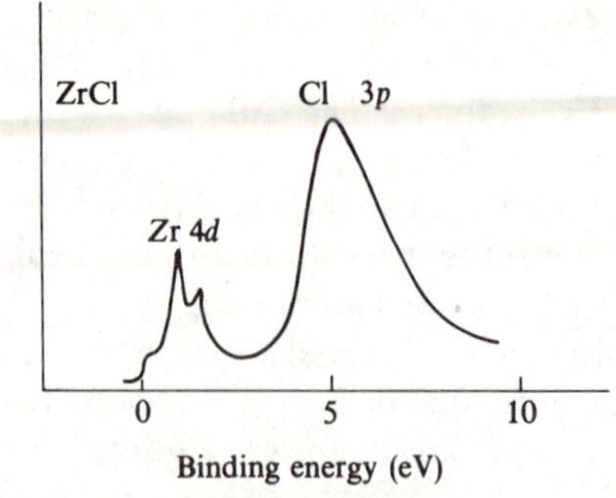

Fig. 3.18 Photoelectron spectrum of ZrCl, showing the Zr 4*d* and the Cl 3*p* bands. (After J. D. Corbett and J. W. Anderegg, *Inorg. Chem.*, **19** (1980), 3822.)

$Mo_6Cl_8^{4+}$ units present in $MoCl_2$, where the metal–metal bonding is confined to discrete clusters of atoms.

3.4.4 The failure of the band picture

The simple band model suggests that any solid with a partially filled band should be metallic. Many compounds with incompletely filled shells of *d* and *f* electrons, however, are non-metallic. In Section 3.4.2 it was shown that such behaviour can sometimes be attributed to a ligand field splitting of the *d* band. In this situation the lower part of the *d* band may be full. There are very many cases, however, where non-metallic solids cannot be understood in these terms. A majority of simple transition-metal halides, as well as many oxides and other compounds, form insulating solids with spectroscopic and magnetic properties characteristic of partially filled *d* shells. The absorption spectrum of nickel oxide, NiO, for example, is very similar to that shown by $Ni(H_2O)_6^{2+}$ in aqueous solution, where the nickel ions are well separated from each other. The magnetic properties of NiO show that each Ni^{2+} ion has two unpaired electrons, as is also found in isolated Ni^{2+} complexes. Thus nickel oxide cannot have a filled band, and according to the band model should be metallic. In fact, pure nickel oxide is a good insulator, although it may have semiconducting properties if it is not stoichiometric. The breakdown of band theory is very common for compounds of transition metals, and even more so for lanthanides. The reason for this failure of the simple model is the repulsion between electrons which has been neglected so far. Electrostatic repulsion tends to keep electrons localized on individual atoms. It is only when the band-forming tendency caused by orbital overlap is strong enough, that metallic properties are observed. The fact that TiO and VO are metallic, but the oxides MnO, FeO, CoO and NiO are not, can be understood from the larger overlap of *d* orbitals earlier in the transition series. This is partly related to the defective structures of TiO and VO referred to earlier, but is chiefly a consequence of the contraction of the *d* orbitals, which occurs as the nuclear charge increases. The 3*d* bandwidth in the later monoxides is not zero, and in fact measurements by PES suggest values of around 1 eV. This is much smaller than in the metallic oxides, however, and is not sufficient to overcome the localizing influence of electron repulsion.

The breakdown of band theory has far-reaching consequences for the properties of many transition-metal and lanthanide compounds. For this reason, it will be discussed in detail in Chapter 5.

Further reading

A good account of the uses and limitations of the ionic model in chemistry is to be found in:

C. S. G. Phillips and R. J. P. Williams (1965). *Inorganic chemistry*, Vol. 1, Chapter 5; [see also Vol. 2, Chapter 31]. Oxford University Press.

Another critical discussion, emphasizing the difficulties involved in defining 'real' ionic character, is:

C. R. A. Catlow and A. M. Stoneham (1983). *J. Phys. C: Solid State Phys.*, **16** 4321.

Many of the topics discussed in the present chapter are also treated in more detail in other texts. The one mentioned below is particularly concerned with covalent bonding in solids, and discusses dielectric properties and the Phillips–van Vechten treatment of ionicity. However, there are also useful discussions of transition metals and transition-metal compounds.

W. A. Harrison (1980). *Electronic structure and the properties of solids.* W. H. Freeman.

Relationships between bonding and structure are also discussed in:

J. C. Phillips (1967). Chemical bonds in solids. In *Treatise on solid state chemistry* (ed. N. B. Hannay) Vol. 1, (Plenu Press).
J. C. Phillips (1970). *Rev. Mod. Phys.* **42**, 317.
D. M. Adams (1974). *Inorganic solids.* John Wiley and Sons.
J. K. Burdett (1979). *Nature*, **279** 121.
J. K. Burdett (1980). *J. Amer. Chem. Soc.*, **102**, 450.

The free-electron theory of metals is described in:

B. R. Coles and A. D. Caplin (1976). *Electronic structures of solids.* Edward Arnold.

C. Kittel (1976). *Introduction to solid state physics* (5th edn), Chapter 6. John Wiley and Sons.

Accounts of the *d* levels and ligand field effects in transition-metal complexes, are to be found in many general books on inorganic chemistry, and in:

D. Nicholls (1974). *Complexes and first-row transition elements.* MacMillan.

For two reviews of solids containing extended metal–metal bonding, see:

A. Simon (1981). *Angew. Chem.: Int. Ed. Engl.* **20** 1.
R. E. McCarley (1981). In *Mixed-valence compounds* (ed. D. B. Brown). D. Riedel.

4
Elementary band theory

The previous chapters have shown how energy bands for electrons in solids arise from overlapping atomic orbitals. At an elementary level, we have seen how it is possible to interpret the bands in various types of solid in terms of simple chemical bonding models. For a more advanced understanding of the electronic properties it is necessary to look in more detail at the form of the **crystal orbitals** which make up the bands. This is the subject of the present chapter. So far as possible, we shall treat the crystal orbitals using the LCAO method familiar to chemists from the theory of molecular orbitals. Nevertheless, it will appear that several new concepts are necessary to deal with the solid state. Although some of these ideas are referred to in subsequent chapters, a detailed understanding of band theory is not really essential for these later parts. Because of the more theoretical nature of the present chapter, it could be omitted on a first reading of the book.

It is necessary at this point to make explicit the assumption that we are dealing with a **crystal**, in which atoms form a regular array. Since solids contain a very large number of atoms, it is only in such regular situations, where the electron distribution is repeated from one unit cell to the next, that simple models are possible. The detailed electronic orbitals of disordered solids are much more difficult to treat: the topic of defects and disorder is considered in Chapter 7.

Real solids are of course three-dimensional. Many of the ideas of band theory, however, can be introduced by considering simpler hypothetical models, in which atoms form arrays in one or two dimensions only. This allows us to picture the form of the crystal orbitals much more easily than in three dimensions, and to understand clearly the consequences of the lattice periodicity. We shall start therefore with the simplest case possible: that is, a single row of atoms, forming a 'one-dimensional crystal'.

4.1 Crystal orbitals in one dimension

4.1.1 The importance of periodicity—Bloch functions

Our problem is to find the approximate form of the wave function for an electron moving along a row of atoms (see Fig. 4.1). Atoms at the ends of the

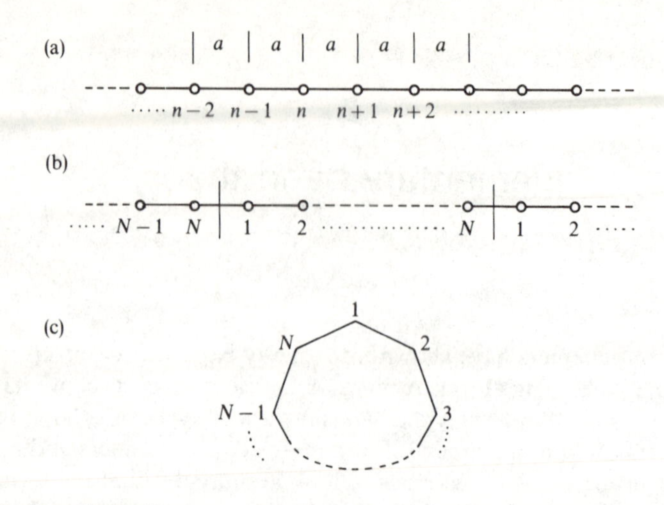

Fig. 4.1 One dimensional monatomic chain. (a) Lattice spacing a and numbering of atoms. (b) Periodic boundary conditions, where the chain of N atoms is assumed to repeat identically in either direction. (c) Alternative view of the meaning of the periodic boundary condition. Atoms 1 and N are linked.

chain represent the surface of the solid, where the electronic structure may be different from the bulk. For the moment, however, it is simpler not to worry about such effects, and to eliminate somehow the ends of the chain. It would be possible to start with an infinite number of atoms, but this creates difficulties, for example when we need to count the number of possible crystal orbitals. A way round this problem is to introduce the fiction of **periodic boundary conditions.** We consider the behaviour of electrons in a finite chain of N atoms, but eliminate surface effects by assuming that the chain is repeated identically an infinite number of times (Fig. 4.1(b)). We shall find that the wave function for the electrons varies along the chain. The periodic boundary condition implies however, that it comes back to exactly the same value after N lattice spacings. Thus if $\psi(x)$ is the wave function along the chain, the result of moving the electron N atoms along gives $\psi(x+Na)$, where a is the lattice spacing. We assume that:

$$\psi(x+Na)=\psi(x). \tag{4.1}$$

Another way of picturing the periodic boundary condition is shown in Fig. 4.1. We could imagine that the N atoms are joined in a ring, rather than forming a straight chain, and that atom number 1 is joined to atom number N.

Although the periodic boundary conditions seem rather artificial when we first encounter them, they allow us in a sense to 'have our cake and eat it': that is, we only need to think about a finite number of atoms, so that the orbitals can be counted without difficulty, but at the same time the ends of the chain are eliminated, so that we do not have to worry about surface effects. At first, it is easier to think of a chain with a small number of atoms. In this case, the periodic boundary conditions have a rather serious effect on the allowed wave functions and their energies. Later, however, we shall derive formulae which do not depend on the number of atoms, and we can then make N as large as we like, so as to simulate a 'real' crystal.

Since the spacing of atoms along the chain is uniform, the wave function must reflect this regularity. If we translate the chain by one atomic spacing, the electron density must be unchanged. In order to satisfy this condition it is necessary to use a wave function which is *complex* and has real and imaginary parts. The electron density (clearly a real function !) is then given by:

$$\rho(x) = \psi^*(x)\psi(x). \tag{4.2}$$

The periodicity of electron density in the chain means that:

$$\rho(x + a) = \rho(x). \tag{4.3}$$

This can be only be achieved if:

$$\psi(x + a) = \mu\psi(x) \tag{4.4}$$

where μ is a complex number such that

$$\mu^*\mu = 1. \tag{4.5}$$

The effect of translation through a number of lattice spacings, say n, gives:

$$\psi(x + na) = \mu^n\psi(x). \tag{4.6}$$

The assumption of periodic boundary conditions, equation 4.1, then implies that μ must satisfy the equation:

$$\mu^N = 1. \tag{4.7}$$

Thus μ must be a complex Nth root of unity. There are N different solutions of equation 4.7, given by the general formula:

$$\mu = \exp(2\pi ip/N) = \cos(2\pi p/N) + i\sin(2\pi p/N) \tag{4.8}$$

where p is an integer and $i = \sqrt{-1}$. p can be regarded as a quantum number, which labels the wave functions. However, it is usual in the theory of solids to define a slightly different quantum number k, which is proportional to p:

$$k = 2\pi p/(Na). \tag{4.9}$$

We shall find that using k, rather than the quantum number p, enables us to write formulae which do not depend on the number of atoms N. For a finite chain, k can only take discrete values, given by putting $p = 0, \pm 1, \pm 2, \ldots$ in equation 4.9. However, in a real solid N is very large, and the difference between successive allowed values of k is very small. Thus in practice we can regard k as a continuous variable.

Returning to equation 4.4, we see that the effect of translating the chain by one lattice spacing is given by:

$$\psi(x+a) = \exp(ika)\psi(x). \tag{4.10}$$

One possible wave function which satisfies this equation is that corresponding to free electron waves (see Fig. 4.2):

$$\psi(x) = \exp(ikx) = \cos(kx) + i\sin(kx). \tag{4.11}$$

However, equation 4.10 allows a much more general form for the wave function:

$$\psi(x) = \exp(ikx)u(x) \tag{4.12}$$

where $u(x)$ is any function which is *periodic*, and is unaltered by moving from one atom to another:

$$u(x+a) = u(x). \tag{4.13}$$

The general form of equation 4.12 is known as a **Bloch function.** For chemical pictures of bonding, it is best to construct Bloch functions from overlapping atomic orbitals, and some examples are shown in Fig. 4.2. The atomic orbitals on each atom now form the periodic function $u(x)$, and the amplitude of the wave function is modulated by the term $\exp(ikx)$ in equation (4.12). We can see that all the functions have a wave-like form, with a wavelength λ determined by the quantum number k. In fact:

$$\lambda = 2\pi/k. \tag{4.14}$$

Figure 4.2 shows three values of k: zero, corresponding to infinite wavelength; a small value giving a large but finite wavelength; and the value π/a, where the wavelength is just two lattice spacings, and atomic orbitals on adjacent atoms are out of phase with one another.

The fundamental result of this section is that the periodic arrangement of atoms forces the wave functions for electrons to satisfy the Bloch function equation (4.12). This equation can be applied to the free-electron model, or to wave functions written as linear combinations of atomic orbitals. The significance of the quantum number k will become clearer in the following sections. We have seen that it determines the wavelength of a crystal orbital, and for this reason, it is called the **wave number**. In the free-electron theory, k is proportional to the momentum of an electron, and is therefore important in treating properties such as conductivity, which depend on the motion of electrons in crystals. We shall see that it is possible to plot the energy as a function of k in the LCAO method, as well as in the free-electron model. k is also an important quantum number in spectroscopy, as it governs the selection rules for electronic transitions between different energy bands.

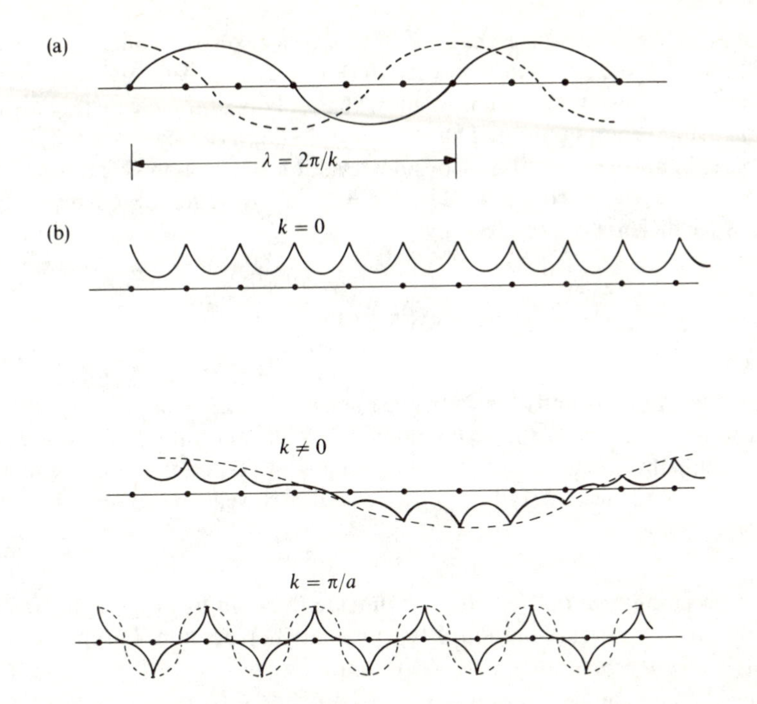

Fig. 4.2 Wave functions for electrons along a chain. (a) Real (———) and imaginary (– – – –) parts of a free-electron wave function. (b) Bloch functions with different k values, formed from overlapping atomic orbitals. Only the real part of the wave function is shown.

4.1.2 LCAO theory of the monatomic chain

Having shown the general form of the wave function for an electron in a periodic lattice, we shall now consider a little more specifically a band composed of overlapping atomic orbitals. We shall apply the **linear combinations of atomic orbitals** (LCAO) approximation, to the chain of atoms shown in Fig. 4.1. We assume at first that each atom has just one valence s orbital. If the wave function for an atomic orbital on atom number n is written $\chi_n(x)$, we write a crystal orbital in the form:

$$\psi(x) = \sum_n c_n \chi_n(x). \tag{4.15}$$

The coefficients c_n can be determined by requiring that our crystal orbital should behave like the Bloch function of equation 4.12. Since the distance x down the chain to atom n is simply na, we find that:

$$c_n = \exp(ikna). \tag{4.16}$$

We have seen that for a chain of N atoms, k takes the values allowed by equation 4.9, $k = 2\pi p/(Na)$, where $p = 0, \pm 1, \pm 2, \ldots$. Although any integral value of p is allowed, we shall now show that only N of them generate distinct crystal orbital combinations. This is not surprising, as we have started from N atomic orbital basis functions. Consider a value k', corresponding to a number $p+N$ in the above equation. Then $k' = k + 2\pi/a$, and the corresponding orbital coefficients c_n' are given by:

$$\begin{aligned} c_n' &= \exp\{i(k+2\pi/a)na\} \\ &= \exp(ikna)\cdot\exp(2\pi in) \\ &= c_n. \end{aligned} \tag{4.17}$$

So the two values, k and k', give the coefficients for the *same* crystal orbital in equation 4.15. A range of $2\pi/a$ contains just N allowed values of k, and values of k outside this range do not give new orbitals. We could take k in the range 0 to $2\pi/a$, but it is usual to allow negative values as well, and to take the range:

$$-\pi/a \leqslant k < +\pi/a. \tag{4.18}$$

We must emphasize that values of k outside these limits do give satisfactory orbital coefficients in equation 4.16, but that they are simply repetitions of orbitals already generated, and so are unnecessary.

The crystal orbitals given by equations 4.15 and 4.16 are called **Bloch sums** of atomic orbitals. They have already been illustrated in Fig. 4.2. It can be seen that the $k = 0$ combinations have all atoms in phase, and so correspond to bonding combinations of low energy. As k moves away from zero, nodes are introduced into the wave function, and the energy will rise. The states of highest energy in the Fig. 4.2 are those with $k = \pi/a$, where all adjacent atoms are combined in an antibonding manner. This argument can be made more quantitative by evaluating the energy of each orbital from:

$$E_k = \frac{\int \psi_k^* \mathscr{H} \psi_k}{\int \psi_k^* \psi_k}. \tag{4.19}$$

$\mathscr{H}$ is the Hamiltonian operator, and we have given a subscript k to the energy, and to each crystal orbital. The next step is to express ψ_k and its complex conjugate ψ_k^* as Bloch sums. We find for the numerator and denominator of equation (4.19):

$$\int \psi_k^* \mathscr{H} \psi_k = \sum_{n=1}^{N} \left\{ \sum_{m=1}^{N} \exp[i(n-m)k] \int \chi_m^* \mathscr{H} \chi_n \right\} \tag{4.20}$$

and

$$\int \psi_k^* \psi_k = \sum_{n=1}^{N} \left\{ \sum_{m=1}^{N} \exp[i(n-m)k] \int \chi_m^* \chi_n \right\}. \tag{4.21}$$

The individual integrals over atomic orbitals can be evaluated by computer programs, but the essential features of the electronic structure can best be shown by an approximate model, similar to the Hückel MO treatment of π-electron systems.

For the integrals in equation 4.21, we assume (a) that each atomic orbital χ_n is normalized, and (b) that the overlap of atomic orbitals on different centres can be neglected. Thus:

$$\int \chi_n^* \chi_n = 1 \tag{4.22}$$

and

$$\int \chi_m^* \chi_n = 0 \quad \text{if } n \neq m. \tag{4.23}$$

The neglect of overlap may seem to be a very serious approximation. It can be justified in a more sophisticated theory, by assuming that we are not starting from genuine atomic orbitals, but from linear combinations of them made beforehand, chosen to eliminate the overlap. With this assumption, only terms with $n = m$ contribute to the sum in equation 4.21. There are N identical terms, each equal to one, so that:

$$\int \psi_k^* \psi_k = N. \tag{4.24}$$

For the Hamiltonian matrix elements in equation 4.20, those with $n = m$ simply give the energy of an electron in one atomic orbital. We denote this α. Terms with $n \neq m$ give the energies of interaction between different atomic orbitals, and it is reasonable to neglect interactions between orbitals that are not on neighbouring atoms in the chain. For the interaction between neighbours, we write β. So we have:

$$\int \chi_n^* \mathscr{H} \chi_n = \alpha \tag{4.25}$$

and

$$\int \chi_m^* \mathscr{H} \chi_n = \beta \text{ if } n \text{ and } m \text{ are neighbours.} \tag{4.26}$$

In counting neighbours we must remember the periodic boundary conditions, which effectively make atoms 1 and N next to each other (see Fig. 4.1, p. 80). Equation 4.20 therefore has N identical terms, each with a contribution for $n = m$, and contributions from the two neighbours of a given atom. Thus:

$$\int \psi_k^* \mathscr{H} \psi_k = N\{\alpha + \beta[\exp(-ika) + \exp(ika)]\} \tag{4.27}$$

$$= N\{\alpha + 2\beta \cos(ka)\}. \tag{4.28}$$

Combining equations 4.24 and 4.28, we find the following very simple equation for the energy of the orbital ψ_k:

$$E_k = \alpha + 2\beta \cos(ka). \qquad (4.29)$$

For s orbitals which overlap in a bonding fashion, the interaction integral β is negative (corresponding to a lowering of energy) and the resulting function $E(k)$ is plotted in the lower part of Fig. 4.3. The total energy span of the band is from $+2\beta$ to -2β. Thus the width is equal to $4|\beta|$, which is proportional to the degree of interaction between neighbouring atoms. Strongly overlapping orbitals give large values of $|\beta|$, and wide bands, whereas contracted atomic orbitals that overlap poorly in the solid give rise to narrow bands. We have used this idea qualitatively in previous chapters. The more quantitative result obtained here can be extended to two- and three-dimensional solids.

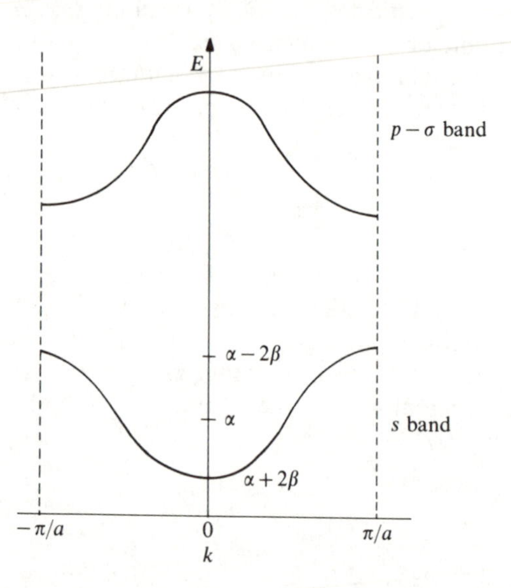

Fig. 4.3 Energy as a function of k for bands of s and $p\sigma$ orbitals in a linear chain.

If the atoms in the chain have other valence atomic orbitals, these can also be used to form crystal orbitals by the Bloch sum method. The upper part of Fig. 4.3 shows another band, coming from the overlap of $p\sigma$ atomic orbitals pointing along the chain. The linear combinations of p orbitals corresponding to $k = 0$ and $k = \pm\pi/a$ are shown in Fig. 4.4. Now the in-phase combinations have antibonding overlap, and the appropriate interaction integral β is positive. Equation 4.29 shows that the band has its maximum energy at $k = 0$, rather than $k = \pm\pi/a$ as for the s orbitals.

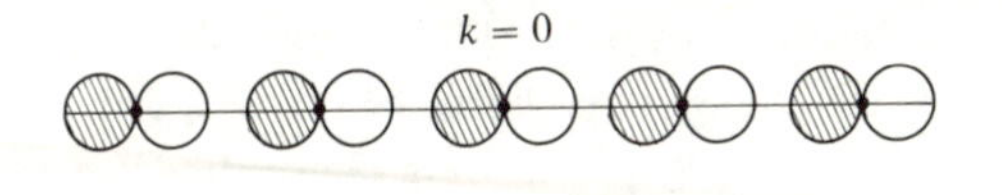

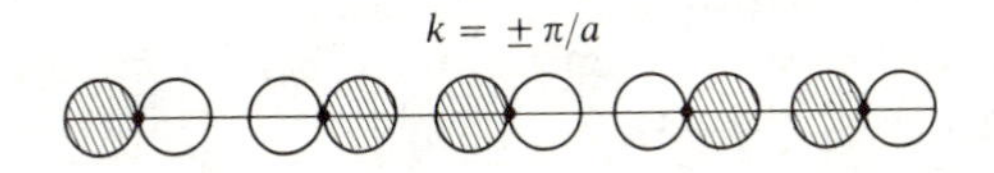

Fig. 4.4 Overlap of $p\sigma$ orbitals for $k = 0$ and $k = \pm\pi/a$.

In general, each atomic orbital can produce a band of N distinct crystal orbitals. In more complex solids, however, the bands formed from different orbitals may overlap in energy. A plot of E against k for the valence and conduction bands in a solid is called a **band structure diagram**. It gives a more detailed view of the energy levels than is contained in the density of states. As we shall see later, the information provided by the band structure is essential for interpreting many of the electronic properties of a solid.

It is interesting to work out the density of states given by the one-dimensional band plotted in Fig. 4.3. There is one orbital for each allowed value of k. From equation 4.9, it can be seen that these values are evenly spaced along the axis of the diagram. This is shown in Fig. 4.5, where the orbital energies resulting from a small number ($N = 8$) of atoms are shown. The energies tend to cluster at the top and bottom of the band. When we make N very large, the allowed values of k are so closely spaced that we can regard them as continuous. The density of states is then found to have infinite peaks at the edges of the band, as shown in Fig. 4.5. This is peculiar to the one-dimensional

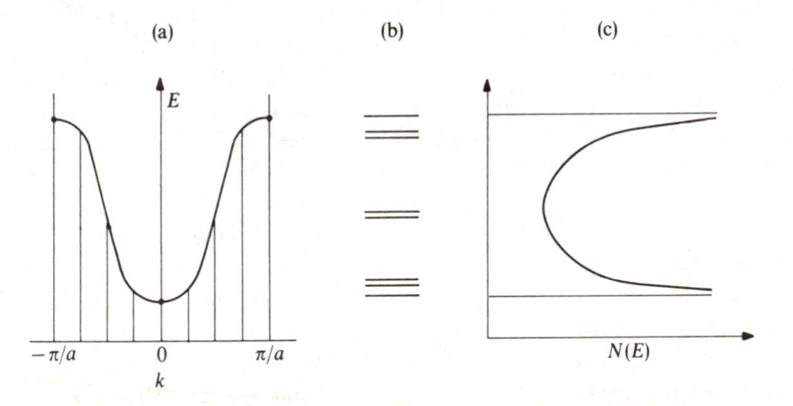

Fig. 4.5 (a) $E(k)$ curve showing the allowed k values for a chain with $N = 8$ atoms. (b) Orbital energies for eight-atom chain, showing clustering at the top and bottom of the band. (c) Density of states for a chain with very large N.

situation, and is quite different from three dimensions, where the density of states is lowest towards the band edges. We can see an approximation to the predicted one-dimensional form in long-chain hydrocarbons. Figure 4.6 shows the photoelectron spectrum of n-$C_{36}H_{74}$, measured at photon energies where the ionization from carbon 2*s* orbitals dominates the spectrum. The strong double-peaked feature comes from a band composed largely of 2*s* orbitals, and has the form shown in Fig. 4.5 for the density of states of a one-dimensional chain.

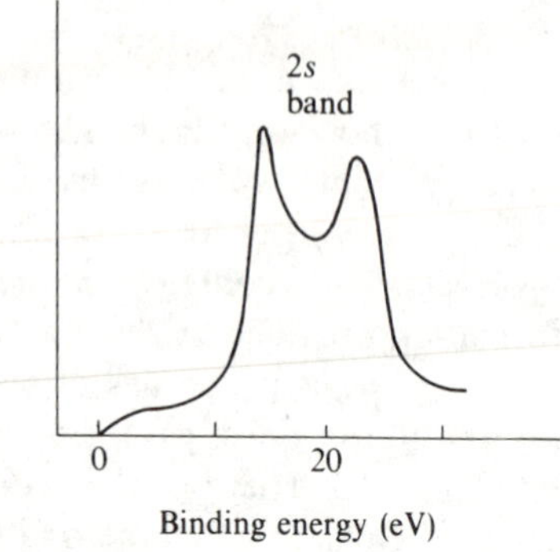

Fig. 4.6 X-ray photoelectron spectrum of the long-chain alkane $C_{36}H_{74}$, showing the density of states in the 2*s* band. (From J. J. Pireau *et al.*, *Phys. Rev. A*, **14** (1976), 2133.)

4.1.3 The binary chain

The monatomic chain was intended as a model of a solid where the predominant electronic interactions are between atoms of the same kind. This may be true in a metallic element, in a molecular solid, or even in an ionic solid where the valence band width comes largely from direct overlap of anion orbitals. In many partially ionic solids however, the most important interactions occur between orbitals on atoms of different kinds. The simplest model of is this situation is provided by a **binary chain**, composed alternately of different atoms A and B (see Fig. 4.7). In the class of solids that we discussed in Section 3.2.2, the most important valence orbitals are the *p* shell on the more electronegative atom B, and the *s* orbital of the less electronegative atom A. To start with, we shall suppose that these are the only atomic orbitals: that is, A has a valence *s* orbital, and B has one valence $p\ \sigma$ orbital, pointing along the chain so as to overlap with A. The unit cell of the lattice, with spacing denoted by a as before, now has two atoms, and two atomic orbitals. Although the unit cell is more complicated, the wave functions must still be Bloch functions. Thus the crystal orbitals can be written as Bloch sums of atomic orbitals as before, but each sum now has a contribution from both of the atoms in the unit

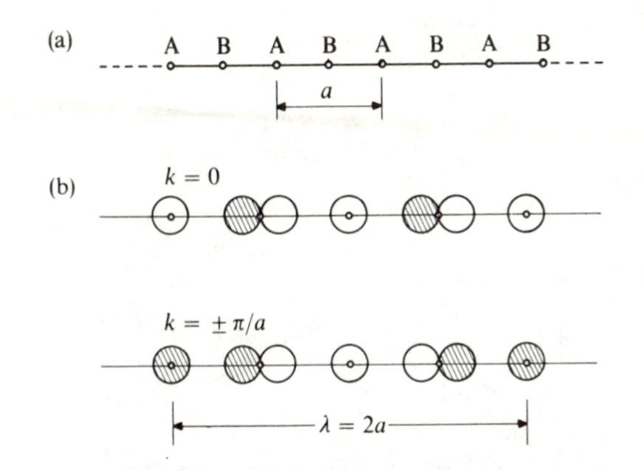

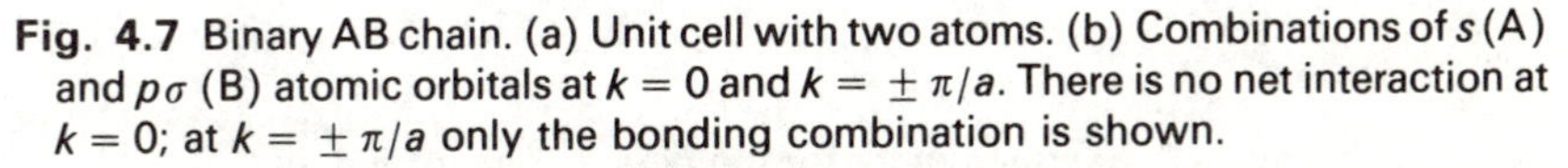

Fig. 4.7 Binary AB chain. (a) Unit cell with two atoms. (b) Combinations of s (A) and $p\sigma$ (B) atomic orbitals at $k = 0$ and $k = \pm\pi/a$. There is no net interaction at $k = 0$; at $k = \pm\pi/a$ only the bonding combination is shown.

cell. Since there are two atomic orbitals, there will be two crystal orbitals for each k value, corresponding to bonding and antibonding combinations.

If the atomic orbitals on A and B at position n in the chain are written $\chi(\mathrm{A})_n$ and $\chi(\mathrm{B})_n$, the bonding and antibonding crystal orbitals can be written:

$$\psi_k = \sum_{n=1}^{N} \exp(ikna)\,[a_k\chi(\mathrm{A})_n + b_k\chi(\mathrm{B})_n] \tag{4.30a}$$

and

$$\psi_k^1 = \sum_{n=1}^{N} \exp(ikna)\,[b_k\chi(\mathrm{A})_n - a_k\chi(\mathrm{B})_n]. \tag{4.30b}$$

a_k and b_k are mixing coefficients, which depend on the relative energies and the degree of overlap between A and B orbitals. The mixing also depends on the wave number k, and the reason for this can be seen in Fig. 4.7(b), where the crystal orbitals are drawn for $k = 0$ and $k = \pi/a$. At $k = 0$ the positive overlap on one side of each atom is exactly cancelled by a negative overlap on the other side. The different symmetry of s and p orbitals therefore means that there is no interaction between A and B at zero wave-vector. We should then put $a = 0$ and $b = 1$ in equation 4.30, and the energies are just those of $\chi(\mathrm{A})$ and $\chi(\mathrm{B})$. At $k = \pi/a$, however, all the overlaps are of the same sign. We now get a bonding combination, of lower energy than $\chi(\mathrm{B})$, and an antibonding one of higher energy than $\chi(\mathrm{A})$. This is just like the molecular situation considered in Section 3.2.2, and illustrated there in Fig. 3.6 (p. 57). The interaction between the orbitals in the AB chain must increase gradually as k moves away from zero, and reaches its maximum value at $k = \pi/a$.

The band structure, or $E(k)$ plot for the binary chain is shown in Fig. 4.8. The atomic composition of the two bands varies with k, although the lower one

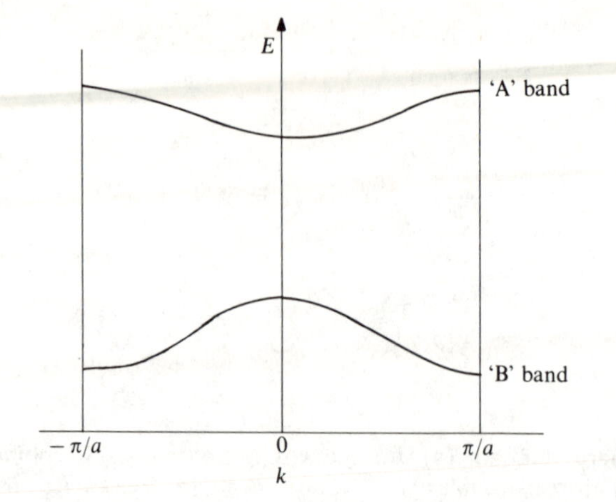

Fig. 4.8 Band structure for the binary chain of Fig. 4.7.

has more B character and the upper one more from A. We can see how the width of the bands is now controlled by the formation of bonding and antibonding orbitals in the heteropolar AB system. The degree of interaction depends on the overlap between A and B orbitals, and inversely on the energy difference between the atomic orbitals. If the energy difference is large, mixing will be small and the bands narrow. This corresponds to the highly ionic situation, where the lower band is almost purely B in composition, and the upper one mostly A.

A more complete band structure for the binary chain is plotted in Fig. 4.9, including bands formed from a set of *s* and *p* orbitals on each of the atoms A and B. The dotted lines show the bands formed from *p*-π orbitals, directed at right angles to the direction of the chain. The reader may like to draw the *p*-π combinations corresponding to different values of *k*, and to verify that the maximum interaction between these occurs at $k = 0$, with zero interaction at $k = \pi/a$. Only one-half of the complete diagram is shown, since crystal orbitals with $+k$ and $-k$ always have the same energy.

The binary chain model shows how bands can arise from the covalent overlap between atomic orbitals of different energy. This is important, for example, in transition-metal compounds such as ReO_3, where metallic character comes from the width of the Re 5*d* bands. The rhenium atoms in this solid are too far apart for direct overlap to be effective, and the main contribution to the bandwidth is from indirect covalent interactions of Re 5*d* and O 2*p* orbitals. Covalent bonding is enhanced by the high formal oxidation state of rhenium, which draws the 5*d* orbitals down in energy, closer to that of the oxygen orbitals.

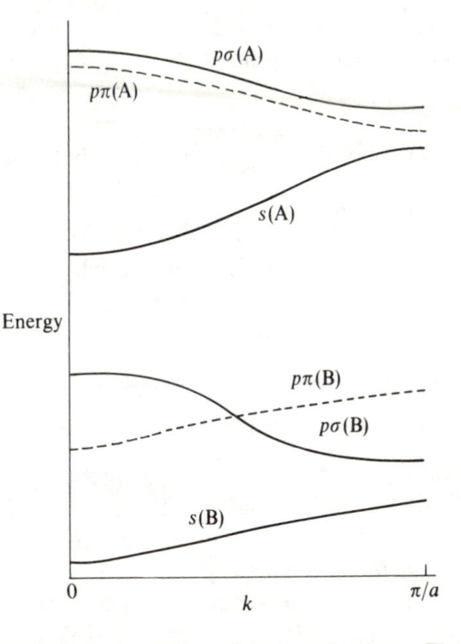

Fig. 4.9 More complete band structure for the AB chain. The principal atomic orbital constituent of each band is indicated.

4.1.4 The nearly-free electron model

The free-electron theory forms an entirely different starting point for the theory of solids. Rather than taking the orbitals of each atom as our basis, we ignore them completely, and begin with the wave functions for electrons moving freely through the solid. We have seen in Chapter 3 that this theory works well for simple metals. Because of its simplicity, the free-electron model is in fact used widely to treat the electronic properties of metals and even semiconductors. We shall look here at the basic model, and show later how it can be modified for wider application.

For electrons moving in a constant potential, the wave function is just that given in equation 4.11:

$$\psi = \exp(ikx)$$

although sometimes it is more convenient to use the real orbitals that can be made from combinations of the degenerate functions with $+k$ and $-k$:

$$\psi_c = \cos(kx)$$
$$\psi_s = \sin(kx).$$

We saw previously that k is related to the wavelength of the electron:

$$\lambda = 2\pi/k.$$

Now for free electrons, we can use de Broglie's formula which relates the wavelength of a particle to its momentum p:

$$p = h/\lambda \tag{4.31}$$

Combining the last two equations, we find that the momentum p is proportional to k:

$$p = \hbar k. \tag{4.32}$$

We can calculate the energy of the electron, which is the sum of the kinetic energy:

$$T = mv^2/2 = p^2/(2m) \tag{4.33}$$

and the constant potential energy, V_0. For a free electron therefore:

$$E(k) = V_0 + p^2/(2m) = \hbar^2 k^2/(2m) + V_0. \tag{4.34}$$

Up to this point, the electron might be in free space, as we have entirely ignored the effect of the periodic lattice. The potential for an electron in a crystal cannot be constant, but must depend on the distance to the nearest atom. In a periodic chain of atoms, the potential might vary in the kind of way suggested in Fig. 4.10(a). It is important to consider the effect of this on the free-electron waves. Figure 4.10 shows the cos (kx) and sin (kx) waves for various values of k in the periodic potential. With small k, where the wavelength is long, the electron merely experiences an average potential, and is unlikely to be much affected by the 'bumps'. However, as k approaches $\pm\pi/a$, the effect becomes stronger, until at $k = \pi/a$, the cosine and sine waves fit exactly into the lattice spacing. The cos (kx) wave function has its maximum electron density at the atomic positions, where the potential is lowest, whereas the sin (kx) function has zero electron density there. The two functions must therefore have different energies.

The energy of the electrons in the periodic potential is shown plotted against k in Fig. 4.11. The dashed curve is the parabola of equation 4.34, appropriate to electrons moving in a constant potential. The effect of the periodic potential is shown by the full line. The splitting at $k = \pm\pi/a$ is the energy difference between the cosine and sine orbitals mentioned above, and we can see that a gap appears in the band at this point. The same argument shows that there should be an energy gap whenever an integral number of half-wavelengths fit into the lattice, and this happens when the wave number is any multiple of $\pm\pi/a$. The diagrams show the electron waves with $k = \pm 2\pi/a$, and the corresponding energy gap in the band structure. We can think of the energy gaps in the free-electron model as being due to the strong *interference* of some electron waves in the lattice, and in fact the corresponding wavelengths are just those that satisfy the Bragg condition for diffraction of electrons by the lattice.

In the chemical picture, band gaps in solids arise from the energy difference between different atomic orbitals, or between bonding and antibonding

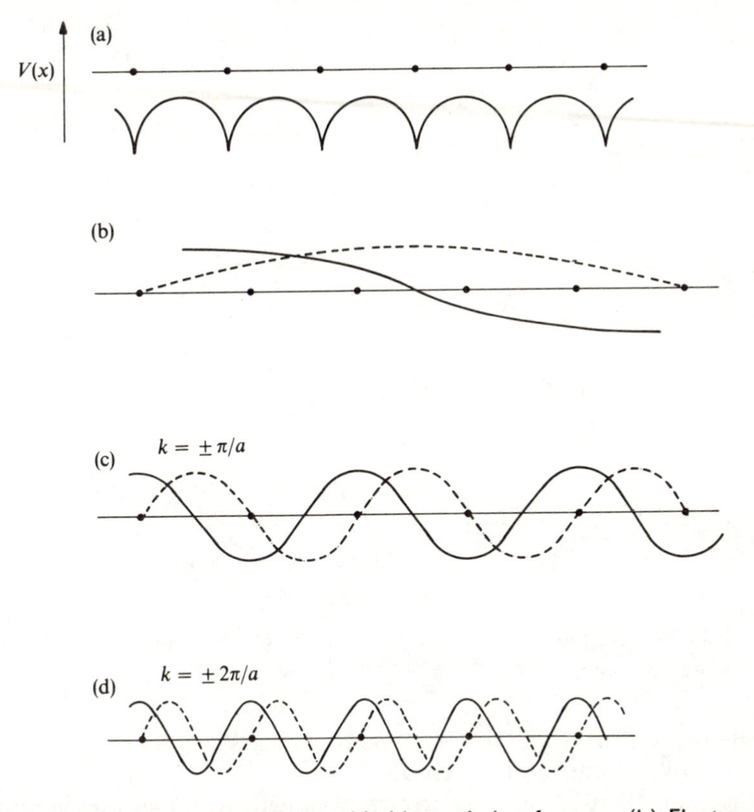

Fig. 4.10 (a) Periodic potential provided by a chain of atoms. (b) Electrons of long wavelength experience an average potential. (c) and (d) Electron waves of appropriate k have sine (---) and cosine (———) components with nodes in regions of different potential.

combinations. The free-electron model gives a very different view. In this picture, band gaps come from the interaction of free electrons with the periodic potential in a lattice. The two approaches are not mutually incompatible, however. It is the potential field of the atoms that provides the periodic lattice potential. In an isolated atom, this same potential field gives rise to atomic orbitals with different energies. The energy bands of a solid are determined by the competition between the potentials of individual atoms, and the overlap of their orbitals. The free-electron model works most simply when the overlap is very strong, so that the atomic orbitals tend to lose their identity. As we have seen, this is the particularly true in the simple metals formed by pre-transition elements. Covalent and even ionic solids however, can be treated from a free-electron starting point, although a much stronger periodic potential is then

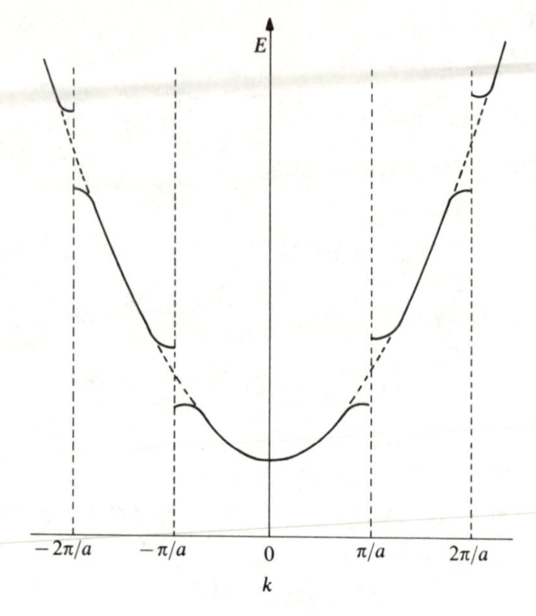

Fig. 4.11 Nearly-free electron-band structure. (----) $E(k)$ with no lattice potential; (——) effect of the periodic potential.

required. From the chemical point of view, it is easiest to think of these solids in terms of overlapping atomic orbitals.

The connection between the free-electron and LCAO approaches can be brought out more clearly by drawing the free-electron band structure in a slightly different way. In this theory the wave number k is proportional to the momentum of the electron, and can have any value that we like. In the LCAO picture on the other hand, k was introduced as a quantum number, and we found it necessary to limit its values to the range $-\pi/a \leqslant k < +\pi/a$, so as not to duplicate the crystal orbitals. Suppose however that we have a free-electron wave function with some particular value of k. It is always possible to write:

$$k = k' + 2n\pi/a \tag{4.35}$$

where k' is in the limited range above, and n is some integer. Then:

$$\begin{aligned}\psi(x) &= \exp(ikx)\\ &= \exp(ik'x)\cdot\exp(2n\pi ix/a).\end{aligned} \tag{4.36}$$

The last factor is a periodic function, as it is unchanged by increasing x by a lattice spacing a. This is precisely the form of a Bloch function, defined in equation (4.12). If we like therefore, we can express any free electron wave as a Bloch function, with a value of k in the range $-\pi/a$ to $+\pi/a$. We could redefine the meaning of k in the free-electron picture, and shift all values by some multiple of $2\pi/a$, so that they lie in this range. The main motivation for this is to plot the free-electron band structure over the same range of k as in the LCAO theory. The result is shown in Fig. 4.12. The replotted diagram shows a remarkable qualitative similarity to the LCAO band structure in Fig. 4.3. (There is a third band in Fig. 4.12, which in the LCAO approach would come from another set of atomic orbitals at higher energy.) The essential similarity between free-electron and LCAO-band diagrams holds in two and three dimensions, although we shall find later that there is an important difference from the one-dimensional case. In the one-dimensional lattice, any periodic potential, however weak, produces energy gaps in the free-electron states. This is not the case in two or three dimensions, where quite strong periodic potentials are necessary to give band gaps.

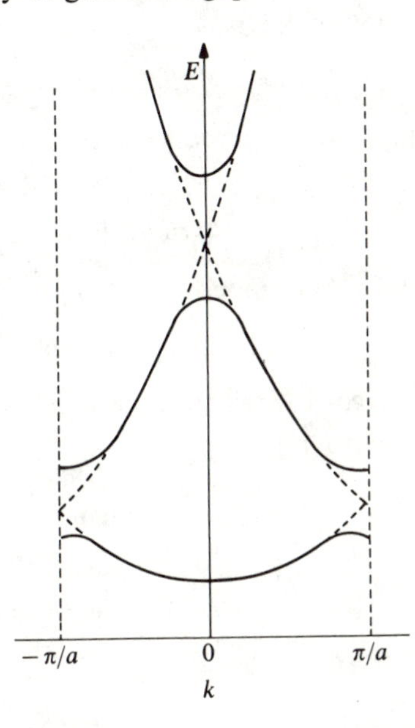

Fig. 4.12 Nearly-free electron-band structure of Fig. 4.11 plotted in the first Brillouin zone.

The range of wave-vectors used to plot the band structure in Figs 4.3 and 4.12 is often called the **First Brillouin Zone**. In order to interpret these diagrams correctly, it is important to realize that the values $+\pi/a$ and $-\pi/a$ correspond to the *same* crystal orbital. This has a most peculiar consequence in the free-electron theory, where k is proportional to momentum. If we accelerate an electron, its momentum will increase until its wave-vector reaches the Brillouin zone boundary, at $+\pi/a$. It will then 'reappear' on the other side of the diagram, with the opposite momentum corresponding to $k = -\pi/a$.

Numerous experiments on metals show that electrons really show this strange behaviour. The simplest physical interpretation uses the diffraction idea mentioned above. The Brillouin zone boundary corresponds to the electron wavelength that satisfies the Bragg condition. An electron with wave-vector $+\pi/a$ can therefore be diffracted into the state where $k = -\pi/a$.

4.1.5 The effective mass concept

It is interesting now to make a slightly more quantitative comparison between the shapes of the $E(k)$ curves predicted by the LCAO and free-electron theories. An electron near the bottom of a free-electron band does not feel much effect of the periodic potential, and the energy is given quite well by the unperturbed formula in equation 4.34. This same parabolic form is shown in the LCAO formula (equation 4.29):

$$E(k) = \alpha + 2\beta \cos(ka).$$

Since we are considering k values close to zero, we can expand $\cos(ka)$ in a power series:

$$\cos(ka) = 1 - (ka)^2/2 + \ldots .$$

Thus for small k in the LCAO model:

$$E(k) = (\alpha + 2\beta) - (ka)^2/\beta. \tag{4.37}$$

On the other hand, the free-electron formula (equation 4.34) is:

$$E(k) = V_0 + (\hbar k)^2/(2m).$$

Remembering that for the s band $\beta < 0$, we can equate these two results if:

$$V_0 = \alpha + 2\beta \tag{4.38}$$

and

$$(\hbar k)^2/(2m) = -\beta(ka)^2. \tag{4.39}$$

The last equation gives:

$$-\beta = \hbar^2/(2ma^2). \tag{4.40}$$

The constant potential V_0 in the free-electron theory can be chosen at will, but on the other hand, if the bottom of the band is to have the same shape in the two models, the interaction between atomic orbitals in the LCAO theory must be related to the interatomic spacing a, as shown in equation 4.40. In simple metals, and indeed in other situations where the bands arise from the strong overlap of valence s and p orbitals, this equation is approximately obeyed. There are many cases however, where the interaction of atomic orbitals does not produce a band of the correct width to agree with the unmodified free-electron theory. Because the model is useful for calculating the physical properties of solids, it is common to 'adjust' it so as to become more widely applicable. This is done by treating the electron mass as a variable parameter, that is by replacing m by an *effective mass*, written m^*. At the bottom of a band, the effective mass is defined by fitting a modified free-electron formula to the actual $E(k)$ curve. We replace equation 4.34 by:

$$E(k) = V_0 + (hk)^2/(2m^*). \tag{4.41}$$

If $E(k)$ is given by the LCAO form, equation 4.40 shows that we must have:

$$m^* = -\hbar^2/(2\beta a^2). \tag{4.42}$$

The electron therefore has a small effective mass in wide bands, where $|\beta|$ is large, and a large effective mass if the band is narrow.

The justification for using the effective mass, as a kind of 'fiddle factor' in the free-electron model, is that it allows many of the simple predictions of the model to be applied more generally than would otherwise be possible. We shall see that electrons with small effective mass do in fact behave as 'light', mobile particles. On the other hand, 'heavy' electrons in narrow bands have lower mobilities, and can be easily trapped by impurities or by distortions of the lattice.

The effective mass can be defined in a more general way. If we differentiate equation 4.41 twice with respect to k, we obtain:

$$d^2E/dk^2 = \hbar^2/m^*, \tag{4.43}$$

or:

$$m^* = \hbar^2/(d^2E/dk^2). \tag{4.44}$$

This shows that the effective mass for electrons at any point in a band depends on the curvature of $E(k)$. Equation 4.44 has rather odd consequences near the top of a band, where d^2E/dk^2, and hence m^*, is negative. The dynamics of electrons can be studied by applying electric or magnetic fields to the solid. For example, in an electric field $\mathscr{E}$ Newton's Second Law of Motion shows that the acceleration of an electron of mass m and charge $-e$ is:

$$d^2x/dt^2 = -(e/m^*)\mathscr{E}. \tag{4.45}$$

Various experiments show that (e/m^*) for an electron near the top of a band does indeed have the opposite sign to that of a free electron. Normally this can only be done with electrons at the Fermi level in a nearly full band; a lone electron rapidly loses energy by exciting lattice vibrations. Rather than thinking of negative effective mass, however, this situation is normally described in slightly different language. The current carriers in nearly full bands are regarded as positively charged *holes*, each having a positive mass. In the next section, we shall show how the notion of positive holes is useful in discussing the conductivity of solids with nearly full bands.

4.1.6 Electronic conductivity

The free-electron model shows that the quantum number k, introduced at the beginning of this chapter as a general consequence of the periodic nature of a crystal, is related to the momentum of an electron, according to the equation:

$$k = \hbar p$$

Plots of the $E(k)$ function are therefore useful for discussing the dynamics of electrons in solids. In this section, we shall derive a formula for the electrical conductivity. Although we are still restricted to motion in one dimension, the result is in fact applicable to two- and three-dimensional solids.

Orbitals with $+k$ and $-k$ have the same energy, and so the ground state of any solid must contain equal numbers of electrons moving in opposite directions. Thus there can be no net motion, and no electric current flowing. This is shown in Fig. 4.13(a) and (b). The heavy line in these $E(k)$ plots indicates the occupied levels in the band. In Fig. 4.13(a), that of a filled band, there is no possibility of having more electrons moving in one direction than in the other, so that no electronic conduction is possible. This corresponds to the normal picture of an insulator, discussed in the Chapter 1. On the other hand, in the partially filled band in Fig. 4.13(b), it is possible to displace the k value of each electron from the ground state, so as to produce a net motion of charge. Suppose that the wave-vector of each electron is shifted by δk from the ground state. Then for *free electrons* the change in velocity (v) is:

$$\delta v = \delta p/m = \hbar \delta k/m. \tag{4.46}$$

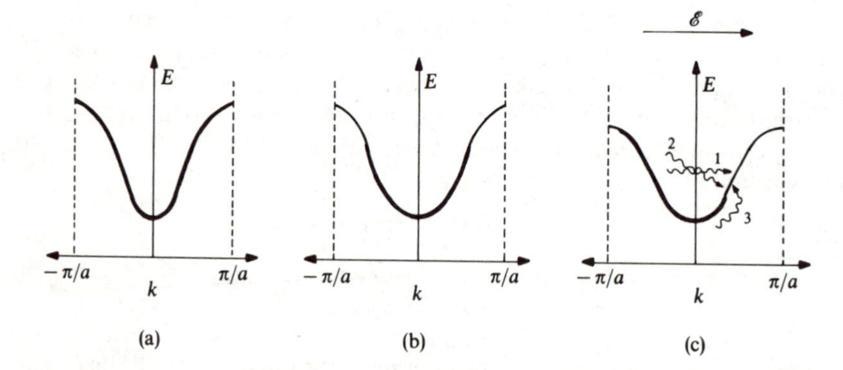

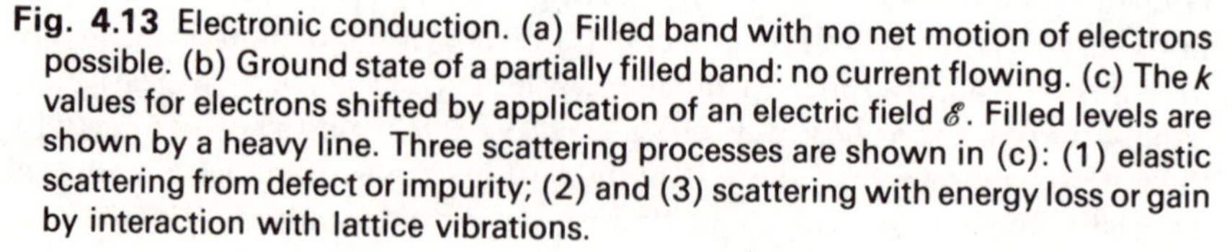

Fig. 4.13 Electronic conduction. (a) Filled band with no net motion of electrons possible. (b) Ground state of a partially filled band: no current flowing. (c) The k values for electrons shifted by application of an electric field $\mathscr{E}$. Filled levels are shown by a heavy line. Three scattering processes are shown in (c): (1) elastic scattering from defect or impurity; (2) and (3) scattering with energy loss or gain by interaction with lattice vibrations.

In a more general case, the electron mass m must be replaced by the effective mass, m^*. The current carried by each electron is its charge multiplied by its velocity. Since there is no net current in the ground state, the current flowing when the k values are shifted can be calculated by adding the velocity changes of all the electrons in the band. If there are n of them, the current (i) is therefore:

$$i = -ne\delta v = -ne\hbar\delta k/m. \tag{4.47}$$

The current-carrying state illustrated in Fig. 4.13(c) has a higher energy than the ground state, because some electrons have been transferred to orbitals higher up in the band. In the absence of an electric field this state would decay rapidly to the ground state in a normal metal. This is because a real solid is not perfectly periodic. The regular crystal lattice is disturbed not only by defects and impurities, but also by lattice vibrations. The Bloch functions with their associated wave numbers k depend however on the assumption of perfect periodicity. The effect of perturbations is to mix up the orbitals with different values of k. In more physical terms, this means that if an electron starts in a state k, it can be scattered into another state, with a different k. Two sorts of scattering process are possible, and are shown in Fig. 4.13(c):

> Scattering from a defect or impurity (represented by 1): this can change k, but not the energy of the electron state.
> Scattering by interaction with lattice vibrations (represented by 2 and 3): this changes the energy of the electron, since a quantum of the lattice vibration is created or destroyed in the process.

Scattering by vibrations increases strongly with temperature, whereas scattering by static impurities or defects is largely independent of temperature.

At a very low temperature in a *superconductor*, the scattering processes are suppressed, but in a normal metal the current will only persist if the electrons are accelerated continuously by an applied electric field. The force on an electron in a field $\mathscr{E}$ is equal to $-e\mathscr{E}$, and from Newton's Second Law:

$$\mathrm{d}p/\mathrm{d}t = -e\mathscr{E}. \tag{4.48}$$

In terms of the wave number, this can be written:

$$\mathrm{d}k/\mathrm{d}t = -e\mathscr{E}/\hbar. \tag{4.49}$$

Suppose that the average time between scattering events for one electron is τ. In this time, the electric field will increment the wave number of each electron by an amount:

$$\delta k = -\tau e\mathscr{E}/\hbar. \tag{4.50}$$

This gives the shift in wave numbers that can be maintained in the presence of the electric field. Equation 4.47 shows that the current (i) flowing in the metal is given by:

$$i = (ne^2\tau/m)\mathscr{E}. \tag{4.51}$$

The proportional relation between field and current is of course Ohm's Law, and therefore the conductivity is given by the simple formula:

$$\sigma = ne^2\tau/m. \tag{4.52}$$

This result is true for free electrons in real three-dimensional metals, and can be used to estimate the scattering time τ. For very pure copper at liquid helium temperatures, the conductivity can be as high as 10^{11} (ohm-cm)$^{-1}$, and this gives τ around 2×10^{-9} s. This seems quite small, but in this time an electron at the Fermi level in copper can travel 0.3 cm. At room temperatures, and for impure specimens, the scattering time and conductivity are likely to be lower by a factor of 10^5.

We derived equation 4.52 using the unmodified free-electron theory. A more general formula can be obtained, simply by replacing the free-electron mass m by an effective mass m^*. One problem, however, is that this simple theory assumes that all electrons have the value of m^*. Since the effective mass depends on the curvature of the $E(k)$ plot, it will in not be the same for all states in the band. Equation 4.52, with m replaced by m^*, is still a useful approximation when there are relatively few electrons near the bottom of a band, since in this region m is nearly constant. When the band is nearly full, however, it certainly does not work. In this case, it is easier use the hole formalism, discussed in the previous section.

Figure 4.14 shows a nearly filled band, both in its ground state, and under the influence of an electric field $\mathscr{E}$. As before, each occupied electron state is moved to a lower (more negative) k value by the field, so that the unoccupied levels

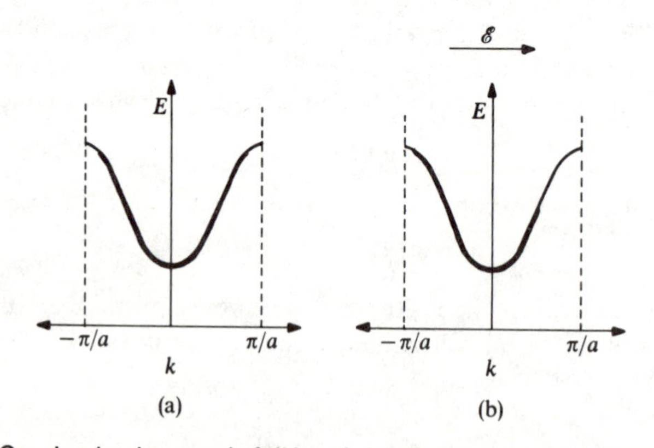

Fig. 4.14 Conduction in a nearly full band. (a) Ground state. (b) Electrons shifted to more negative k values by application of an electric field $\mathscr{E}$: the holes move to more positive k values.

move to higher (more positive) values. We regard each unfilled level as a hole, with positive charge, and a positive mass m_h, given in terms of the effective mass of the missing electron by:

$$m_h = -m^*. \tag{4.53}$$

In the average scattering time τ, each hole is accelerated to give a wave-number change:

$$\delta k = +\tau e\mathscr{E}/\hbar \tag{4.54}$$

On the other hand, the current (i) carried by p holes, each with this k, is:

$$i = +pe\hbar\delta k/m_h = (pe^2\tau/m_h)\mathscr{E}. \tag{4.55}$$

The conductivity provided by the holes is therefore given by a formula quite similar to that for electrons:

$$\sigma = pe^2\tau/m_h. \tag{4.56}$$

The importance of this result is that it shows that the conductivity of a solid with a nearly full band does not depend directly on the electrons present, but rather on the number of unfilled levels, the holes in the band.

Both equations 4.52 and 4.56 relate the conductivity to the number of charge carriers in the solid. However, measurements of conductivity alone cannot give this number, since there are other important factors—the effective mass of carriers, and the frequency with which they are scattered by vibrations or imperfections. It is common to separate out these factors, by defining the *mobility*, μ, of electrons, related to the conductivity by:

$$\sigma = ne\mu. \tag{4.57}$$

Equation 4.52 shows that the mobility of electrons in the free-electron theory is:

$$\mu = e\tau/m \tag{4.58}$$

Similar equations can be written for holes in a nearly filled band. We shall see later that it is possible to obtain independent estimates of the mobilities and carrier concentrations in solids.

Mobilities of electrons and holes depend both on the purity of a solid and on its temperature. Raising the temperature increases the number of thermally excited lattice vibrations, so that electrons are scattered more efficiently. In a metal, the conductivity declines with increasing temperature, due to the decreased mobility. Changing the temperature in a semiconductor, however, may have much more effect on the carrier concentration than on the mobility, and the conductivity usually increases with temperature.

4.1.7 Band structure and spectroscopy

The optical properties of a non-metallic solid, as we have seen in previous chapters, depend on electronic transitions between the valence and conduction bands. We shall now show that spectroscopic transitions in crystals are governed by an important selection rule, which is a consequence of the lattice periodicity and the Bloch functions which make up the crystal orbitals.

In the LCAO model of the monatomic chain, we plotted the bands resulting from valence s and p orbitals (see Fig. 4.3). Let us now imagine that the s band is full, and the p band empty, and that we wish to excite an electron from one band to the other. The spectroscopic transition probability between an initial orbital ψ_i and a final orbital ψ_f depends on the integral

$$\int \psi_f^* \hat{\mu} \mathscr{E} \psi_i, \tag{4.59}$$

where $\mathscr{E}$ is the electric vector of the radiation, and μ the dipole moment operator. In the crystal, our initial and final orbitals will be Bloch functions corresponding to the appropriate bands, and the transition integral can be expanded in terms of atomic orbitals. However, it is easier to show the orbitals pictorially, as in Fig. 4.15. The wavelength of visible or UV radiation is very much longer than the lattice spacing in crystals, and therefore the electric field of the radiation will be very nearly in phase on neighbouring atoms. In Fig. 4.15 are shown the transition dipoles on each atom, determined by the overlap of the initial and final crystal orbitals. For the transition from the $k = 0$ state of the s band, to $k = 0$ in the p band, the atomic transition dipoles are all in phase, and so will combine with the electric field to give a non-zero value for the transition integral (equation 4.59). The transition to a different k state however gives transition dipoles on each atom which are all of different phases, and will cancel out. Figure 4.15(c) shows a transition from a non-zero k value in the s band, to the same k value in the p band. The transition dipoles are now in phase again.

The argument has been expressed rather qualitatively above, but its consequence is clear: transitions are only allowed between Bloch functions with the same wave number, k. That is, we have the selection rule:

$$\Delta k = 0. \tag{4.60}$$

On the $E(k)$ band-structure diagrams therefore, the only transitions that are allowed are ones vertically upwards, or downwards.

The k selection rule expressed in equation 4.60 can be given a different physical significance, using the relation between wave number and momentum in the free-electron theory. Because they have a comparatively very long wavelength, visible or UV photons have a very small momentum. Thus equation 4.60 expresses a momentum conservation law, and implies that the momentum of the electron undergoing the transition may not change.

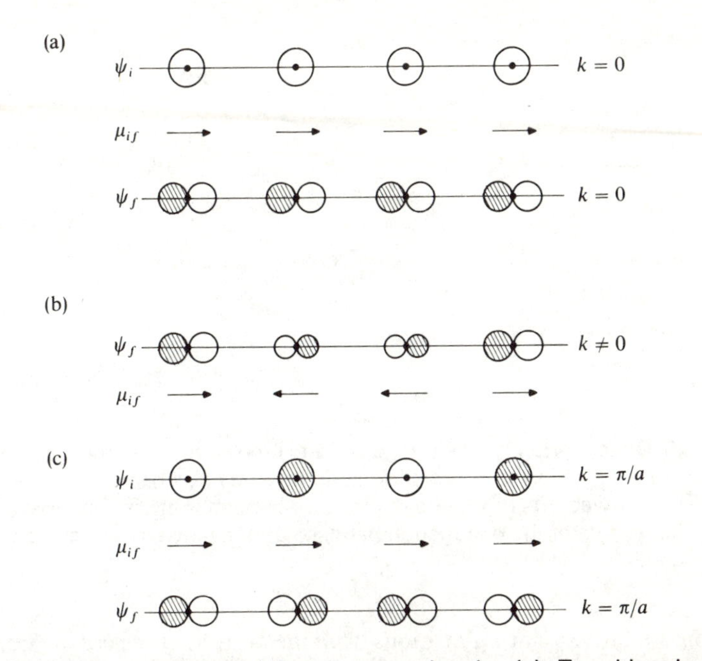

Fig. 4.15 Spectroscopic transitions between bands. (a) Transition between $k = 0$ states of an *s* and a *p* band, showing transition dipoles (μ) in phase. (b) Transition from same *s* state to $k \neq 0$ state of *p* band: transition dipoles now out of phase. (c) Transition between two states with the same non-zero k: transition dipoles again in phase.

The selection rule for electronic transitions will be used later in this chapter to show how $E(k)$ curves can be measured directly by photoelectron spectroscopy. For the moment, however, we can see that this rule may have important consequences for the optical absorption spectrum of a solid. Figure 4.16 shows two different situations that may arise in transitions from a valence band to a conduction band. The band structure in Fig. 4.16(a) shows a case where the top of the valence band and the bottom of the conduction band correspond to orbitals with the same value of k. (It need not be zero, although this is the commonest situation.) In this solid, the lowest energy transition between the bands is allowed. This is known as a **direct band gap**, and is the commonest case for ionic and covalent insulators. In Fig. 4.16(b) however, the lowest energy transition requires a change of wave number, and so is forbidden. This is an **indirect band gap**. Such indirect gaps occur, for example in silicon and in silver bromide (AgBr), and their (three-dimensional) band structures will be dealt with later in this chapter. The same selection rule applies of course to emission spectra, in which electrons come down from excited states into lower unfilled levels. An indirect band gap tends to inhibit the recombination of

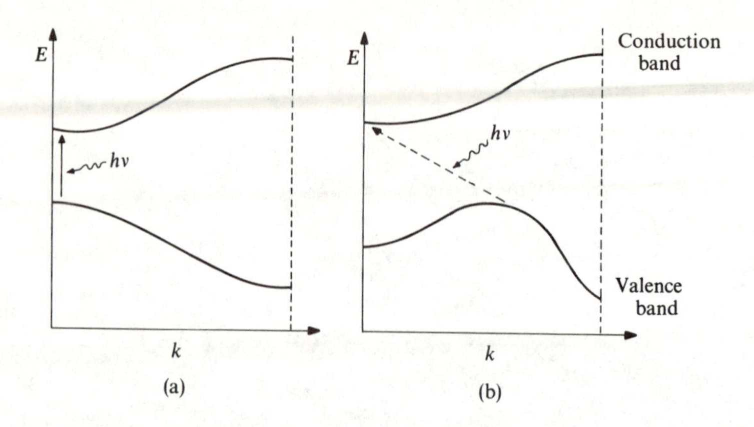

Fig. 4.16 Direct and indirect band gaps. In (a) the lowest energy transition from the valence band to the conduction band involves no change in wave-vector, and so is allowed. In (b) the top of the valence band is at different wave-vector from the bottom of the conduction band, so that the lowest energy transition is forbidden.

electrons and holes, since transitions from the bottom of the conduction band into the top of the valence band are forbidden. The forbidden transitions are observed, although more weakly than allowed ones. The selection rule breaks down for the same reason as in molecules and complexes. For example *d–d* transitions, which are formally forbidden in centro-symmetric complexes, occur because of vibrations that remove the centre of symmetry. In a similar way, the k selection rule depends on perfect lattice periodicity, and is broken by lattice vibrations which destroy this periodicity. Nevertheless, the k selection rule gives a guide to the most intense electronic transitions.

4.2 Two dimensions

Two-dimensional models form a useful 'bridge' between the simple one-dimensional models just considered, and the three-dimensional structure of real solids. At the same time, however, there are many important examples of solids with layer structures, where by far the strongest interaction is between atoms within one layer. In such solids, it is a good first approximation to use a two-dimensional approach, and to neglect the interactions between layers. As with one dimension, we shall start with the LCAO approach, and compare it later with the free-electron model.

4.2.1 LCAO theory of the square lattice

The simplest possible two-dimensional crystal consists of atoms of the same kind arranged on a square lattice, with spacing a. This is illustrated in Fig. 4.17.

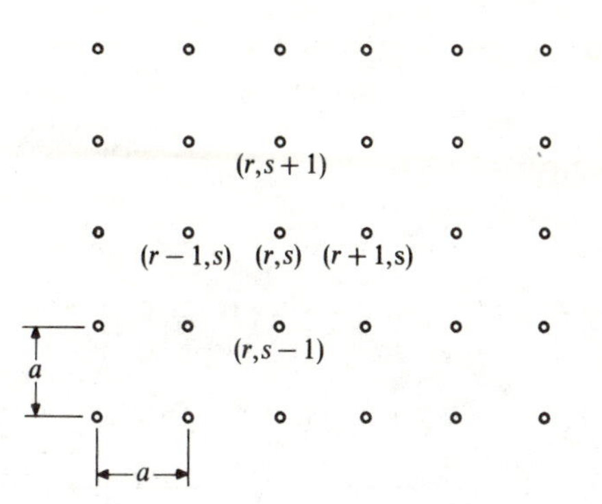

Fig. 4.17 Square lattice, illustrating the labelling of points.

The diagram shows the labelling scheme we shall use for atoms: in two dimensions we need a pair of numbers (r, s), which indicate the position of an atom with respect to the x and y directions.

As with the linear chain, we shall first suppose that each atom has just one valence s orbital, denoted by $\chi_{r,s}$. In the LCAO approach, we need to find crystal orbitals of the form:

$$\psi = \sum_{r,s} c_{r,s}\chi_{r,s} \tag{4.61}$$

The atomic orbital coefficients $c_{r,s}$ are determined as before by the periodicity of the lattice, to give a Bloch sum which is a simple generalization of the one-dimensional case:

$$c_{r,s} = \exp(irk_x a + isk_y a). \tag{4.62}$$

Just as each atom needs two numbers to locate it, so each Bloch sum needs two quantum numbers, (k_x, k_y) to specify its behaviour in the lattice. Figure 4.18 shows the real part of orbitals with different values of (k_x, k_y). It can be seen

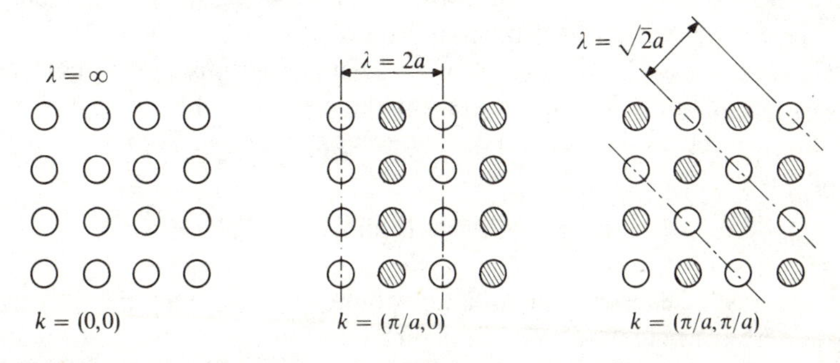

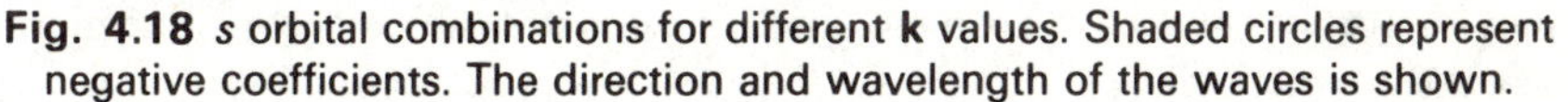

Fig. 4.18 s orbital combinations for different **k** values. Shaded circles represent negative coefficients. The direction and wavelength of the waves is shown.

that the orbitals have a wave-like form in two dimensions, and that the k values show the direction and length of the waves. We call $\mathbf{k} = (k_x, k_y)$ the **wave-vector** of an electron. It points in the direction of greatest change in the wave function, and its magnitude gives the wavelength λ, according to the relation:

$$\lambda = 2\pi/|k| = 2\pi/(k_x^2 + k_y^2)^{1/2}. \tag{4.63}$$

If we have a square array with N atoms in each direction, then just as in one dimension, $\mathbf{k}$ can take the values:

$$(k_x, k_y) = (2\pi/Na)(p, q), \tag{4.64}$$

where p and q are integers. These equations are essentially similar to those in one dimension, discussed in Section 4.1.2. The only difference is that $\mathbf{k}$ is a vector in two dimensions, rather than a simple number. It can be seen from equation 4.62 that changing k_x or k_y by $2\pi/a$ does not change the orbital coefficients, and so we can generate just N^2 different crystal orbitals from our N^2 atomic orbitals. So as not to duplicate the crystal orbitals, we limit $\mathbf{k}$ to the range:

$$-\pi/a \leqslant k_x, k_y < +\pi/a. \tag{4.65}$$

Again, this is just like equation (4.18), which gives the range of k values in one dimension.

Figure 4.19(b) shows that the range of wave-vectors given by equation 4.65 represents a square centred on the origin. This range of $\mathbf{k}$ is the **first Brillouin zone** for the square lattice. Its physical significance in the LCAO model is simply that it gives the range of $\mathbf{k}$ values necessary to generate all possible crystal orbitals from a given set of atomic orbitals, without duplication. The shape of the Brillouin zone depends on the type of lattice, and is sometimes rather less obvious than in the case of the simple square lattice. A general method for constructing the first Brillouin zone depends on the mathematical theory of the **reciprocal lattice**, and is given in Appendix B.

Given the form of the crystal orbitals, it is possible to calculate their energy, just as in one dimension. It is obvious from Fig. 4.18 that the lowest energy orbital for the s band has $\mathbf{k} = (0, 0)$, where all combinations are in phase. As $\mathbf{k}$ increases in magnitude, nodes are introduced and the energy rises. A calculation using the Hückel approximation explained in Section 4.1.2 gives the expression:

$$E(\mathbf{k}) = \alpha + 2\beta\{\cos(k_x a) + \cos(k_y a)\} \tag{4.66}$$

As before, α is the energy of the isolated atomic orbitals, and β the interaction integral between neighbouring atomic orbitals. Only the four atoms in near-neighbour positions have been included, but it is not difficult to include more distant interactions.

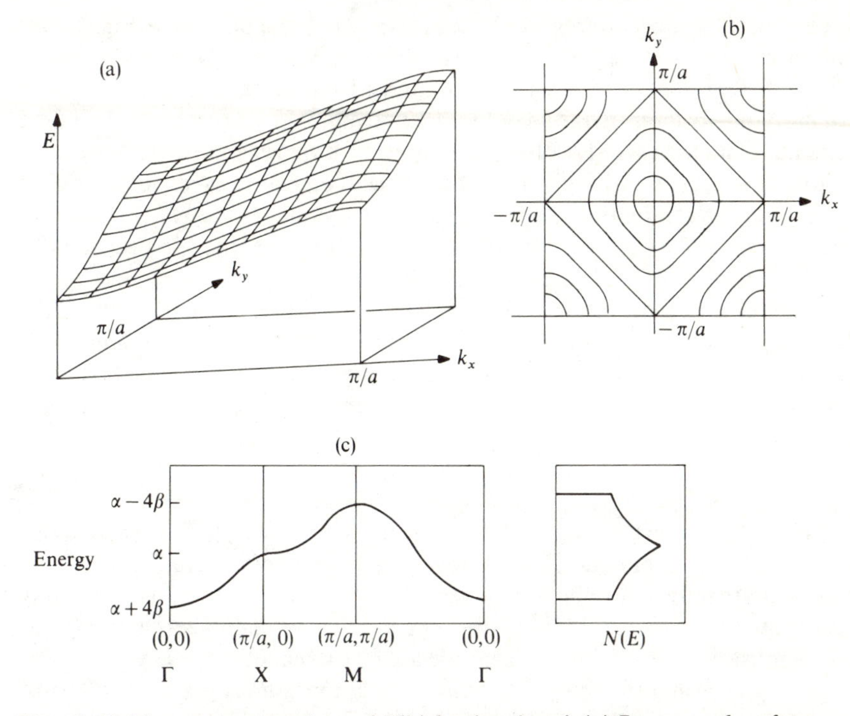

Fig. 4.19 Three representations of $E(\mathbf{k})$ for the *s* band. (a) Energy surface for one-quarter of the Brillouin zone. (b) Constant-energy contours, illustrating the symmetry of the zone. (c) Energy plotted over a triangular path of **k** values, showing minimum and maximum energies, and density of states.

Since **k** is now a two-dimensional vector, it is not so easy to display the $E(\mathbf{k})$ functions as it was in one dimension. Three possible representations are shown in Fig. 4.19. Figure 4.19(a) shows a computer-generated picture of the function in equation 4.66. Only one-quarter of the Brillouin zone is shown, since the other quarters are identical. Figure 4.19(b) shows this symmetry better, by plotting energy contours over the whole zone. Such contour plots are quite useful, but it is not easy to show several bands in this way, and the contour method cannot be used easily in three dimensions. The commonest way of plotting band structures is therefore to take a selection of **k** values, along lines in the Brillouin zone. This is done in Fig. 4.19(c), which shows the energy, first as **k** varies from (0, 0) to $(\pi/a, 0)$, then from $(\pi/a, 0)$ to $(\pi/a, \pi/a)$, and then back to (0, 0). In band theory, the different values of **k** at the centre and edges of the zone are given labels, rather like those for the irreducible representations which arise from the symmetry of molecules. The symmetry properties of a crystal lattice are described by a **space group**, rather than the point groups of isolated molecules. Fortunately, we shall not need to develop the detailed theory of space groups here. It is best to regard the labels shown at the bottom of Fig. 4.19(c) simply as a short-hand notation for the different **k** values. Thus in the square lattice, Γ represents (0, 0), X is (π/a, 0) and M the value (π/a, π/a).

The band-structure diagram of Fig. 4.19(c) cannot give a complete plot of the energies of all the crystal orbitals. The range of **k** values is chosen however, so as to give a representative picture of the range of orbital energies, including the maximum and minimum values. Because of the symmetry of the $E(\mathbf{k})$ function, the energies in the triangle of **k** values shown in the diagram are repeated eight times throughout the allowed range of **k**.

It can be seen that for the *s* band in the square lattice, the total width of the band is now equal to $8|\beta|$. This is double the value found in one dimension, because each atom now has four, rather than two, near neighbours. In a quite general case, where each atom has z near neighbours, the total band width is predicted to be:

$$W = 2z|\beta|. \tag{4.67}$$

Figure 4.19(c) also shows the density of states for the band. The shape of this is quite different from that of one-dimensional bands, plotted in Fig. 4.5. For a finite square lattice, the allowed values of **k** form a series of points, uniformly distributed over the Brillouin zones. The number of energy levels in a given range must now be calculated from the area between adjacent $E(\mathbf{k})$ contours, which is not straightforward. The finite steps at the band edges are also shown in the free-electron model for two dimensions, and are discussed again in the following section.

It is possible to extend the theory just given for the *s* band to allow for more valence orbitals. Figure 4.20, for example, shows Bloch sums formed from p_x

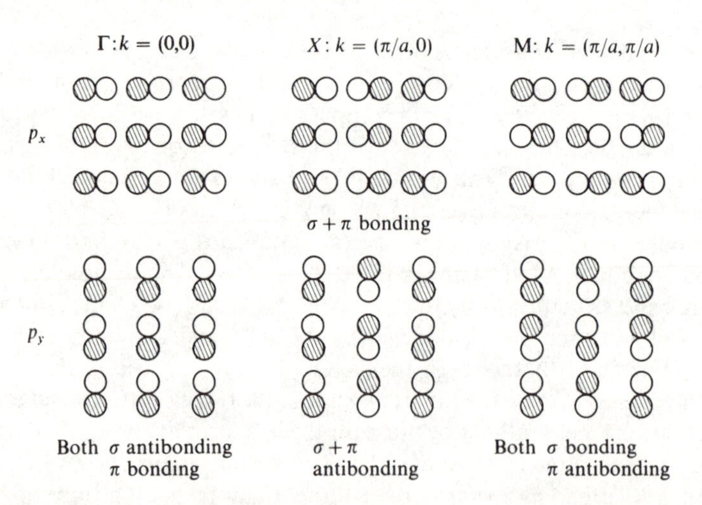

Fig. 4.20 Bloch sums of p_x and p_y orbitals. Shading represents negative lobes.

and p_y orbitals, at three different **k** values. It can be seen from the diagram that at $\mathbf{k} = (0, 0)$ and at $\mathbf{k} = (\pi/a, \pi/a)$, the functions formed from p_x and p_y are degenerate, as they have the same interactions with neighbours. This does not happen at other **k** values, however, such as $(\pi/a, 0)$. The energy now depends on both σ and π interactions between the p orbitals. The band structure diagram, showing $E(\mathbf{k})$ over the same range as before, is plotted in Fig. 4.21.

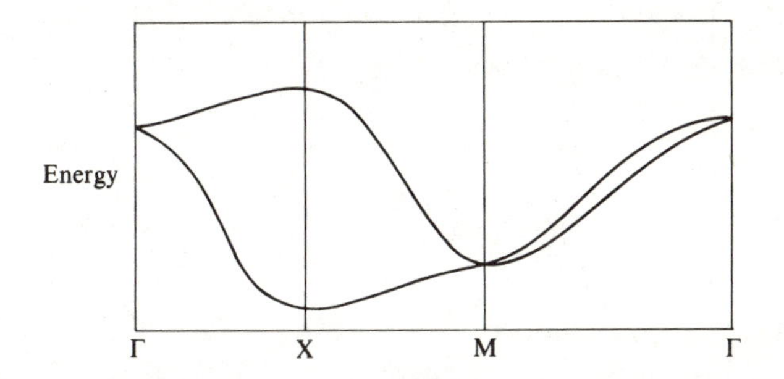

Fig. 4.21 Band structure for the p_x and p_y basis of Fig. 20. The energy is plotted over the same path of **k** values as in Fig. 19.

4.2.2 Nearly free electrons

Like the LCAO theory, the free-electron model in two dimensions may be treated as an extension of the one-dimensional case discussed in Section 4.1.3. The wave functions for the electrons are labelled by a vector **k** with components (k_x, k_y):

$$\psi_{\mathbf{k}}(x, y) = \exp(ik_x x + ik_y y). \tag{4.68}$$

The wave-vector **k** is proportional to the momentum **p** of the electron, and gives its direction as well as its magnitude. Thus:

$$(p_x, p_y) = \hbar(k_x, k_y). \tag{4.69}$$

For electrons of long wavelength, or small momentum, the effect of the periodic potential in the crystal lattice can be neglected. Then the formula for the energy of electrons moving in an average potential V_0 is:

$$\begin{aligned} E &= V_0 + (p_x^2 + p_y^2)/(2m) \\ &= V_0 + (\hbar^2/2m)(k_x^2 + k_y^2). \end{aligned} \tag{4.70}$$

The $E(\mathbf{k})$ function for free electrons is represented in Fig. 4.22: diagram (a) shows a computer-drawn surface, and diagram (b) the circular energy contours on the (k_x, k_y) plane.

We now consider the effect of the periodic potential provided by atoms in a square lattice. In the one-dimensional case, we saw that the interaction with

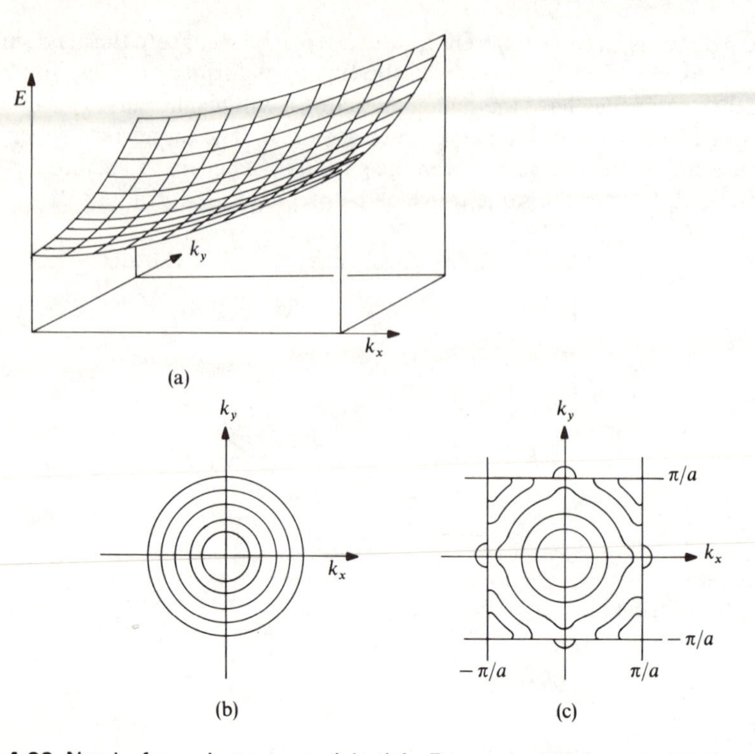

Fig. 4.22 Nearly-free electron model. (a) Energy surface, and (b) energy contours with no periodic potential. (c) Effect of a weak periodic potential on the energy contours.

electrons is most important for values of k where the waves just 'fit into' the lattice: that is when the lattice spacing a is a whole number of half wavelengths. The same is true in two dimensions. When either k_x or k_y is equal to $\pm\pi/a$, the sine- and cosine-like waves have a different energy, exactly as in the one-dimensional case pictured in Fig. 4.10. These are the values of **k** at the Brillouin zone boundary, and as shown in Fig. 4.22(c), the strong interaction of electrons with the lattice gives an energy discontinuity, and distorts the energy contours.

The next step in plotting the band structure is to redefine the free electron wave-vector, exactly as for one dimension. Instead of allowing **k** to have an unlimited range of values, we shift k_x and k_y each by an appropriate multiple of $2\pi/a$, so that all the **k** values fall in the first Brillouin zone. It is now possible to plot $E(\mathbf{k})$ over the same path of **k** values in the zone as for the LCAO energies in the previous section. Figure 4.23(a) shows the resulting curves for free electrons with no periodic potential. The lowest parabolic curves come from small values of momentum, where **k** was originally in the zone. The higher energy bands are states with higher momentum, for which **k** has been shifted

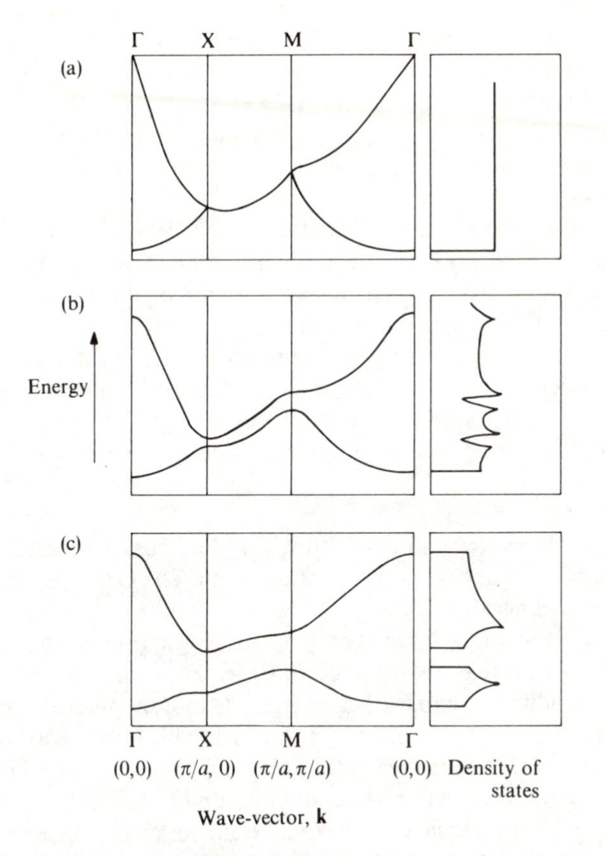

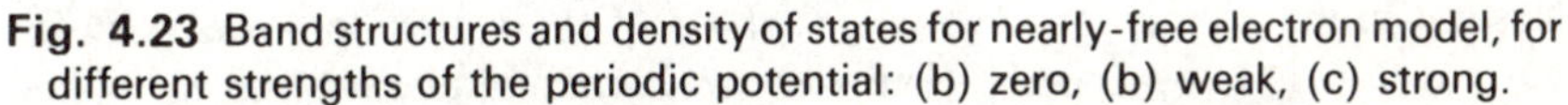
Fig. 4.23 Band structures and density of states for nearly-free electron model, for different strengths of the periodic potential: (b) zero, (b) weak, (c) strong.

into the zone. The density of states for free electrons in two dimensions is also plotted. In Chapter 3 (Section 3.3.1), the density of states for free electrons in three dimensions was derived. Exactly the same argument may be applied to the two-dimensional case, and the interested reader may like to show that the resulting $N(E)$ curve is constant above the bottom of the band, as shown in the diagram.

The effect of a weak periodic potential is shown in Fig. 4.23(b). It can be seen that energy splittings occur at the Brillouin zone edge, just as in one dimension. In the one-dimensional case even a weak periodic potential produces a gap between the bands, but this does not happen in two dimensions. This is due to the extra freedom of motion now possible: although electrons moving in a given direction always have a gap in $E(\mathbf{k})$ when they reach the Brillouin zone edge, this happens at different energies for electrons travelling in different directions in the lattice. A band-gap *can* be produced by a sufficiently strong periodic potential in two (or three) dimensions, however, and this is illustrated in Fig. 4.23(c). The lowest band in the diagram is now very similar to the LCAO s band, and the similarities between LCAO and free-electron models can be treated in a similar way to the one-dimensional model. For example, the free-electron theory can be adjusted by replacing the free-electron mass by an effective mass.

Even for a weak potential which does not produce a band gap, the energy discontinuities in the $E(\mathbf{k})$ curves cause a distortion in the density of states. This is shown in Fig. 4.23(b). When the wave-vector of the electron touches the Brillouin zone edge, there is a pile-up of energy states, giving a peak in the density of states curve, with a sharp drop above it. The same is found for the nearly-free electron model in three dimensions. As we shall see in Section 4.3.2, this feature has been used to rationalize the changes in crystal structure that occur in metallic alloys, as the number of valence electrons is altered.

4.2.3 The Hall effect

The motion of electrons in an electric field has been discussed in the one-dimensional free-electron model (Section 4.1.6). The extra degree of freedom present in two dimensions allows us to see what happens when magnetic field is also applied to the solid. Charged particles moving at right angles to a magnetic field experience a side-ways force, and this gives rise to the **Hall effect**. Consider a conducting slab, with a magnetic field B_z applied at right angles, as in Fig. 4.24. If a current j_x is set up by an electric field along the x direction, each electron tries to follow a curved path. However, if no current is allowed to flow in the y direction, there will be a rapid build-up of charge on the edges of the slab, causing an electric field $\mathscr{E}_y$ just sufficient to keep the electrons moving straight in the x direction. The potential across the y direction caused by the current flow is known as the **Hall voltage**. Its magnitude may be calculated simply for free electrons as follows:

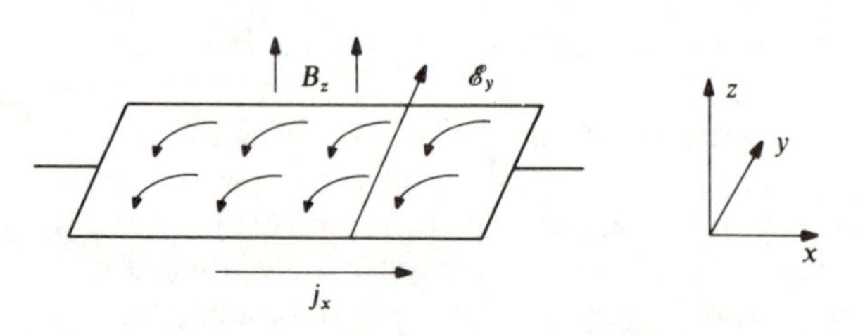

Fig. 4.24 Geometry of fields in the Hall effect. The curved arrows show the initial path of the electrons in the magnetic field (B_z). This leads to a rapid build-up of charge on the edges of the specimen, and an electric field $\mathscr{E}_y$ as shown.

If the current j_x is due to a concentration n of electrons with charge $-e$, the average velocity (v) of each electron must be:

$$v_x = -j_x/(ne). \tag{4.71}$$

The force on an electron (F_y) caused by the magnetic field B_z is:

$$\begin{aligned} F_y &= ev_x B_z \\ &= -j_x B_z/n. \end{aligned} \tag{4.72}$$

When the electrons are constrained to move only in the x direction, the magnetic force must be balanced by an electric field in the y direction, equal to:

$$\begin{aligned} \mathscr{E}_y &= F_y/e \\ &= -j_x B_z/(ne) \end{aligned} \tag{4.73}$$

The ratio:

$$R_{\mathrm{H}} = \mathscr{E}_y/j_x B_z \tag{4.74}$$

is known as the **Hall coefficient** (R_{H}), and for free electrons has the value:

$$R_{\mathrm{H}} = -1/(ne). \tag{4.75}$$

This argument could be repeated for holes in a nearly full band, and we would find:

$$R_{\mathrm{H}} = +1/(pe), \tag{4.76}$$

where p is the hole concentration.

The inverse dependence of R_{H} on the concentration n or p is not difficult to understand. If a given current is carried by fewer charges, each electron or hole must move correspondingly faster, and thus experience a greater force in the magnetic field.

The formulae for the Hall coefficient are valid for three-dimensional solids, since motion of electrons in the z direction in Fig. 4.24 would not give any additional force. Most monovalent metals, such as sodium and copper, have negative Hall coefficients, with values in fairly good agreement with the predictions of the free-electron theory. In many other cases, however, serious differences are found. For example, beryllium and aluminium both have *positive* coefficients, showing that the current is carried more by holes than by electrons. These cases need very elaborate calculations to treat properly, but it is clear that the periodic potential must have an important influence on the motion of electrons.

The most important applications of the Hall effect are to semiconductors where charge carriers are introduced by doping. We can see from equations 4.75 and 4.76 that the carrier type (electron or hole) and concentration can be estimated from the sign and magnitude of the Hall coefficient. If the conductivity is also known, the carrier mobilities, defined in Section 4.1.6, can then be found as well.

4.2.4 The electronic structure of graphite

The electronic levels of solids with layer structures are dominated by the overlap of atomic orbitals within a layer, and it is often a good approximation to ignore the interaction between layers. In this section, we shall look at the electronic structure of graphite, one of the simplest such solids. Figure 4.25 shows how each carbon atom is joined to three others in the graphite layer. σ-bonds between neighbouring atoms are formed from the s orbital and the in-plane p_x and p_y orbitals on each atom. There is consequently a σ-bonding band, and an empty antibonding band, just as in the tetrahedral diamond lattice, discussed in Section 3.2.1 in Chapter 3. The interesting electronic properties of graphite arise from the fourth valence electron on each carbon, which occupies a p_z orbital directed perpendicular to the graphite plane, and is involved in π bonding. We shall now look in detail at the crystal orbitals that form the π bands.

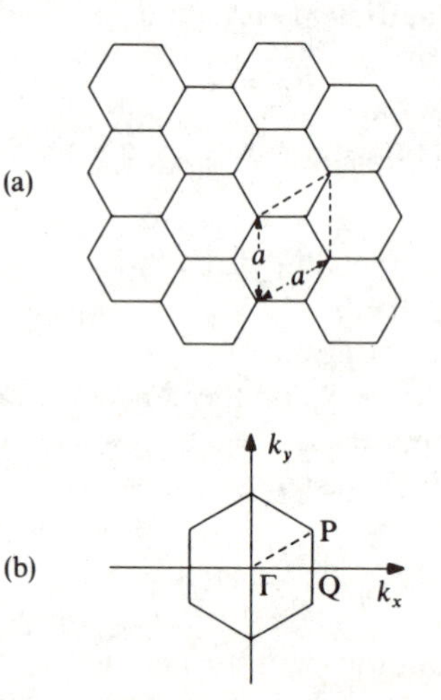

Fig. 4.25(a) Structure of a single graphite layer, showing the unit cell containing two atoms. (b) Brillouin zone for the hexagonal lattice, with the labelling of points Γ, P and Q.

As the Fig. 4.25 shows, the unit cell of the graphite plane contains two atoms. Although they are equivalent by symmetry, the two atoms are not equivalent in terms of the purely translational operations that make up the lattice. Just as with the binary chain of atoms (see Section 4.1.3), the Bloch sums of atomic orbitals must have a contribution from both atoms. For each **k** value, there are therefore two possible combinations, and in most cases one is predominantly

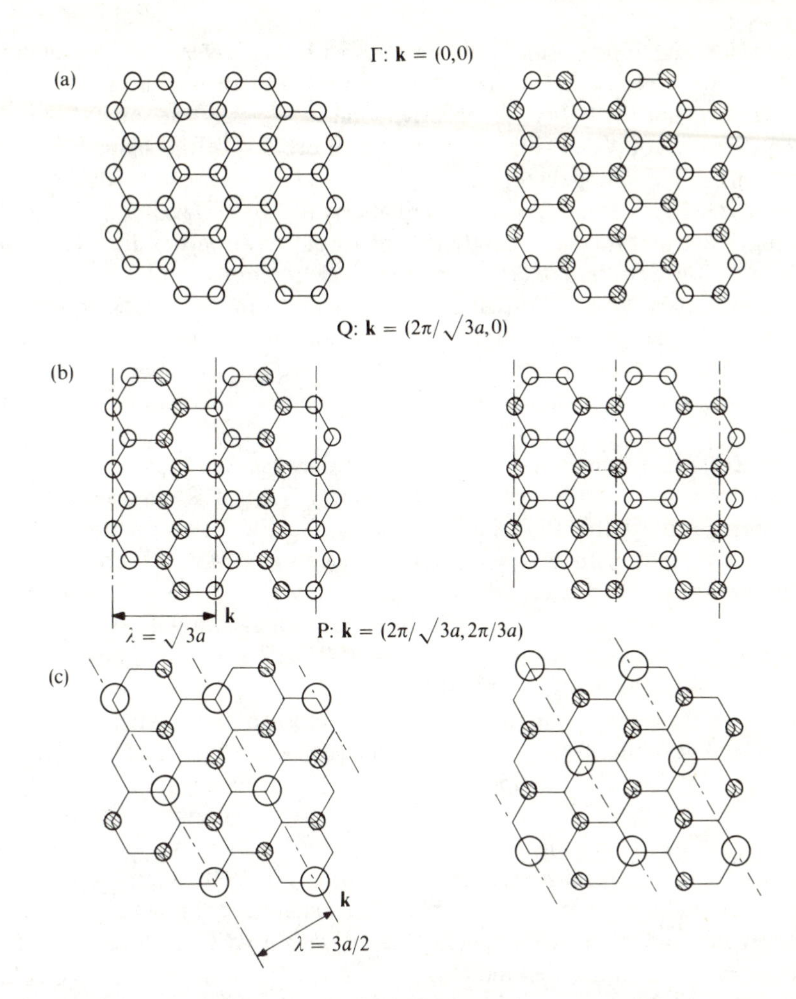

Fig. 4.26 Bloch sums of p_z orbitals for different wave-vectors in graphite. Negative LCAO coefficients are shown shaded circles. At P the non-zero coefficients have the values $+1$ and $-1/2$.

bonding, and the other antibonding, between neighbouring atoms. Figure 4.26 shows some of them. The two $\mathbf{k} = (0, 0)$ orbitals are fully bonding or antibonding, and give the minimum and maximum possible energies in the π band.

As $\mathbf{k}$ increases along the x direction, we start to introduce nodes into the wave function, corresponding to waves in this direction. Figure 4.26(b) shows the bonding and antibonding combinations at $\mathbf{k} = (2\pi/\sqrt{3}a, 0)$. In the lower energy orbital, each atom now has two bonding neighbours and one antibonding, the situation being reversed in the higher energy orbital. This value of **k** is at the edge of the Brillouin zone, with a wave-vector corresponding to the point Q in Fig. 4.25(b). There are six different **k** values, equivalent to Q in the symmetry of the lattice. They correspond to waves of the same length, in symmetrically equivalent directions, and give orbitals of the same energy.

The other interesting direction for **k** is at 30° to the x-axis, towards the point marked P in Fig. 4.25(b). Orbital combinations at P are shown in Fig. 4.26(c). It is remarkable that these two combinations are entirely non-bonding. Each one is located on a different set of non-equivalent atoms, and has no near-neighbour interactions. This is due to the form of the LCAO coefficients at this **k** value. If we try to make bonding and antibonding combinations of the two crystal orbitals at P, we find that each atom has neighbours with orbital coefficients $+1$, $-1/2$, and $-1/2$. The net bonding interaction is therefore zero.

The $E(\mathbf{k})$ values found in the triangle of **k** values marked PQ in the Brillouin zone are repeated twelve times by symmetry, so that it is only necessary to consider the area within this triangle. Figure 4.27 shows two representations of the band structure. In Fig. 4.27(a), the $E(\mathbf{k})$ surfaces for the bonding and antibonding bands are given, showing clearly how the two bands just touch at the non-bonding point P. Figure 4.27(b) is a more conventional band structure plot, taking the energy as **k** goes from $\Gamma = (0, 0)$, to P and to Q. The density of states shows again the touching of the bands.

Each of the π bands, bonding and antibonding, can hold two electrons per unit cell of graphite. Graphite has one electron per carbon atom, so that the lower band is full, and the upper one empty. We can see from the density of states curve that graphite is in a rather unusual class of solids, where there is no energy gap, but where the density of states at the Fermi level is nevertheless zero. It is called a **semi-metal**. In fact, the interaction between the layers in graphite does have some effect, and produces a slight overlap of the two bands. Nevertheless, graphite is a poor conductor, with a very much lower carrier concentration than in most metals.

Graphite forms an interesting series of **intercalation compounds**, in which atoms or molecules are introduced between the layers. In these compounds, there seems to be a high degree of charge transfer between the guest species and the graphite host. Both electropositive and electronegative species can be introduced, typical compounds being LiC_6 and C_8Br. In the former case, the alkali metal gives up an electron, thus populating the lower part of the antibonding π band. With electronegative elements such as Br, electrons are removed from the lower band. Molecules with high electron affinities, such as AsF_5 and PtF_6, may also be intercalated. Most of these intercalation compounds have electrical conductivities higher than that of pure graphite, since there are now many more electronic carriers, either electrons in the upper π band, or holes in the lower one.

The compound boron nitride, BN, is isoelectronic with graphite, and can

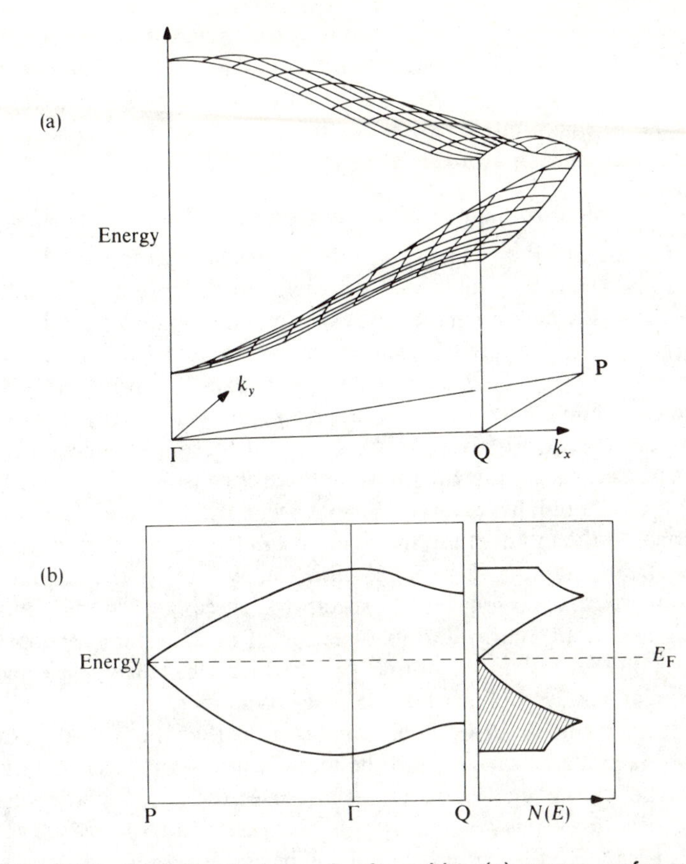

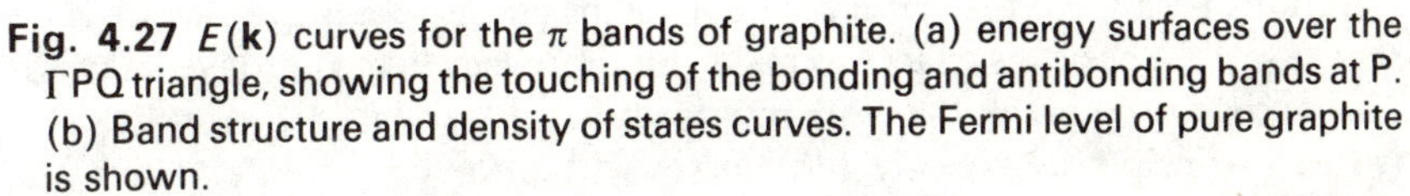

Fig. 4.27 $E(\mathbf{k})$ **curves for the π bands of graphite. (a) energy surfaces over the ΓPQ triangle, showing the touching of the bonding and antibonding bands at P. (b) Band structure and density of states curves. The Fermi level of pure graphite is shown.**

also form a hexagonal layer structure. The band structure resembles that of graphite, except that the lower band has more contribution from nitrogen, which has greater electronegativity. The upper π band has correspondingly more boron contribution. The major difference from graphite is in the orbitals at the non-bonding point P. In BN one of the combinations shown in Fig. 4.26(c) is all on nitrogen atoms, and the other combination all on boron. Because of the energy difference between the two sets of atomic orbitals, the two crystal orbital combinations no longer have the same energy. Thus in BN the upper and lower π bands do not touch, and the solid is an insulator. Electrons in the lower band are more tightly bound than in graphite, and it is therefore much more difficult to form intercalation compounds where electrons are removed. In fact the molecule $(SO_3F)_2$, which has a very high electron affinity, can be intercalated into boron nitride, but this is the only such compound known. The antibonding band is at higher energy than in graphite, and is correspondingly more difficult to populate. Probably for this reason, it is not possible to make intercalation compounds of boron nitride with electropositive species such as alkali metals.

4.2.5 Measurement of band structures by photoelectron spectroscopy

Up to this point, the $E(\mathbf{k})$ curves have been regarded as theoretical constructions which, although they are useful in interpreting the electronic properties of solids, have no immediate experimental significance. It is possible, however, to use photoelectron spectroscopy to measure $E(\mathbf{k})$ curves directly, and thus to verify the predictions of band theory. The principles of these measurements can be understood quite simply from two important properties of the wave-vector $\mathbf{k}$. Firstly, we have seen that there is a selection rule for electronic transitions in the solid. In two or three dimensions where $\mathbf{k}$ is a vector, equation 4.60 should be written $\Delta\mathbf{k} = \mathbf{0}$. Secondly, for free electrons, $\mathbf{k}$ is proportional to the momentum. In the photoelectron experiment, an electron is ionized from the crystal, and becomes totally free in the vacuum. The momentum of an electron in the spectrometer can easily be found, by measuring its kinetic energy and its direction of motion. Since the measurement relies on determining the direction at which electrons are ejected, it is called **angular resolved** photoelectron spectroscopy.

The difficulty with a three-dimensional solid is that the periodicity of the lattice is broken at the surface, and the momentum of the electron emerging from the solid may change. However, in most cases, the surface plane will retain the same *two-dimensional* periodicity as the bulk, and the wave-vector in directions parallel to the surface should still be conserved. In solids with layer structures, the easiest surfaces to make are those parallel to the layers (as it is only necessary to break the weak interlayer bonds), and it is then straightforward in principle to measure $E(\mathbf{k})$ curves within the layers.

The geometrical arrangement is shown in Fig. 4.28. The photoelectron spectrum is measured for electrons emerging at a particular angle (θ) from the

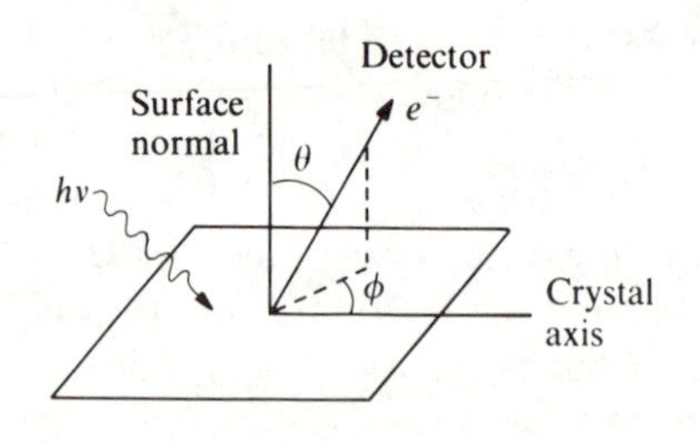

Fig. 4.28 The geometry of the angular-resolved photoelectron spectroscopy experiment, defining the angles θ and ϕ used in the text.

solid. The kinetic energy of the electron (T) gives its total momentum p:

$$p = (2mT)^{1/2}, \tag{4.77}$$

and hence the component of momentum parallel to the surface:

$$\begin{aligned} p_{\parallel} &= p \sin \theta \\ &= (2mT)^{1/2} \sin \theta. \end{aligned} \tag{4.78}$$

The **k** vector of the electron parallel to the surface plane is therefore:

$$\begin{aligned} k_{\parallel} &= p_{\parallel}/\hbar \\ &= (2mT/\hbar^2)^{1/2} \sin \theta. \end{aligned} \tag{4.79}$$

If the electron gets out of the solid without being scattered, this must be the **k** value for the high-energy state to which it was excited by the photon. According to the **k** selection rule, the filled orbital that the electron started from must have the same **k** value. As the angle θ is varied, therefore, the photoelectron spectrum should change in a way that reflects the $E(\mathbf{k})$ curves of the filled bands in the solid. By also changing the polar angle, ϕ in Fig. 4.28, it is possible to plot $E(k)$ curves for different directions of **k** in the crystal lattice.

The spectra of graphite shown in Fig. 4.29 illustrate this technique. For photoelectrons normal to the surface, $k = 0$, and the photoelectron peaks show the energies of orbitals at the centre of the Brillouin zone, denoted by Γ in the previous section. Spectra taken at 30° to the normal are also shown, with **k** in directions corresponding to ΓP to ΓQ. The experimental points shown in Fig. 4.29(b) have been plotted against the **k** values deduced from equation 4.79, in these two directions. The dotted lines show the σ and π bands predicted by theory. The splitting of the π band into two is a result of interaction between the graphite layers. This splitting is quite small however, and the general shape of the π band is very similar to that predicted for a single layer. The experimental results also show that the bottom of the π band and the top of the σ band overlap in energy. This agrees with the X-ray emission spectra, discussed in Section 2.2.2 (see Fig. 2.8).

Unfortunately, the analysis of angular-resolved photoelectron spectra is not always so simple. The form of the high-energy conduction bands, into which the electron is excited before reaching the surface of the solid, sometimes gives additional peaks in the spectrum. There may be features coming from surface states with energies different from the bulk bands. The spectra also have a background due to electrons that have either undergone transitions breaking the **k** selection rule (because of lattice vibrations), or have been scattered before emerging from the solid. In spite of these difficulties, band structure measurements have been made for a large number of solids. As we have seen with graphite, it is possible to make a more direct comparison of theory and experiment than with other techniques.

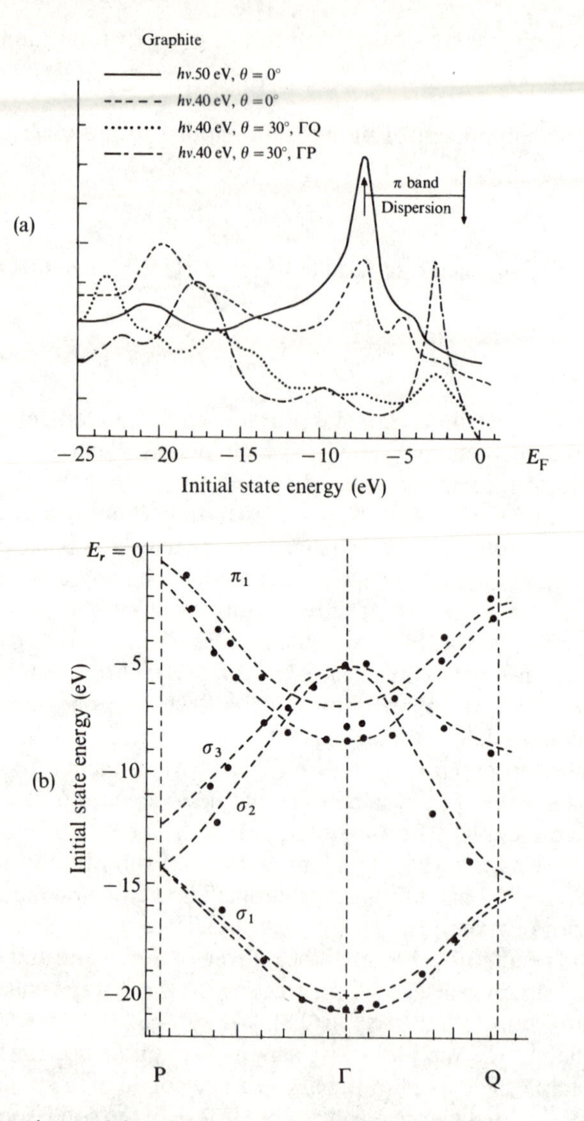

Fig. 4.29 Experimental band structure of graphite. (a) Photoelectron spectra measured at different angles. (b) E(k) curves deduced from the spectra. (From I. T. McGovern *et al.*, *Physica B*, **99** (1980), 415.)

4.3 Three-dimensional band structures

The basic concepts required for three-dimensional solids are essentially the same as in two dimensions. Each crystal orbital is a Bloch function associated with a three-dimensional wave-vector **k**, giving the direction and length of the electron waves in the lattice. The main differences are that it is more difficult to picture the crystal orbitals, and the first Brillouin zone—the range of **k** values necessary to generate all distinct orbitals in a band—is now a three-dimensional region, bounded by planes. The theory of Brillouin zones is treated in Appendix B, and is not really necessary for an appreciation of the applications of band theory.

4.3.1 Band structures of some simple solids

Figures 4.30–4.33 show band structures for some of the solids that we have already discussed from a more chemical point of view. As in two dimensions, the diagrams show the energies of the different bands as a function of the wave-vector, **k**. The labels on the horizontal axis refer to different directions of **k**. For example, Γ is the point **k** = (0,0,0) where all atomic orbitals are in phase, and X refers to **k** in the [100] direction in the lattice.

Figure 4.30 shows a comparison between potassium and silver chloride. The lower bands in KCl are made of combinations of chlorine 3*s* and 3*p* orbitals. These filled valence bands are very narrow, and show hardly any change of energy with **k**. This reflects the small overlap between the atomic orbitals. Using the ideas developed in one dimension, we can see that the electrons in this band will have a high effective mass. In fact holes in the valence bands of alkali halides are completely trapped in the lattice, as we shall see in Chapter 7. The conduction bands are rather broader, because the unfilled potassium orbitals which make them up are more diffuse, and overlap more strongly. The lowest part of the conduction band, at Γ, is composed of 4*s* orbitals, but there are also potassium 3*d* orbitals not much higher in energy, making the details of the band rather complicated. KCl has a direct band gap (see Section 4.1.7) since the top of the valence band is also at Γ. The lowest energy transition is therefore allowed by the *k* selection rule.

In silver chloride the band gap is smaller. The lower energy separation between chlorine and silver valence orbitals also gives rise to a larger degree of covalent mixing between anion and cation orbitals, and this is probably one reason for the greater width of the bands, as compared with KCl. However, there is another very important difference between the two compounds. In AgCl the 4*d* orbitals are full, and occur in the same energy region as the valence band. There is quite strong mixing between the silver 4*d* and the chlorine 3*p* orbitals, which also makes the valence band wider. As illustrated in Fig. 4.31 the different symmetry of *p* and *d* orbitals means that there is no interaction between them when $k = 0$, because of the cancelling between bonding and antibonding overlaps. Mixing is possible however for non-zero wave-vectors, and the formation of bonding and antibonding combinations forces the top of the valence band up in energy. As can be seen in the band structure, the lowest point of the conduction band occurs at Γ as in KCl, but in AgCl the highest

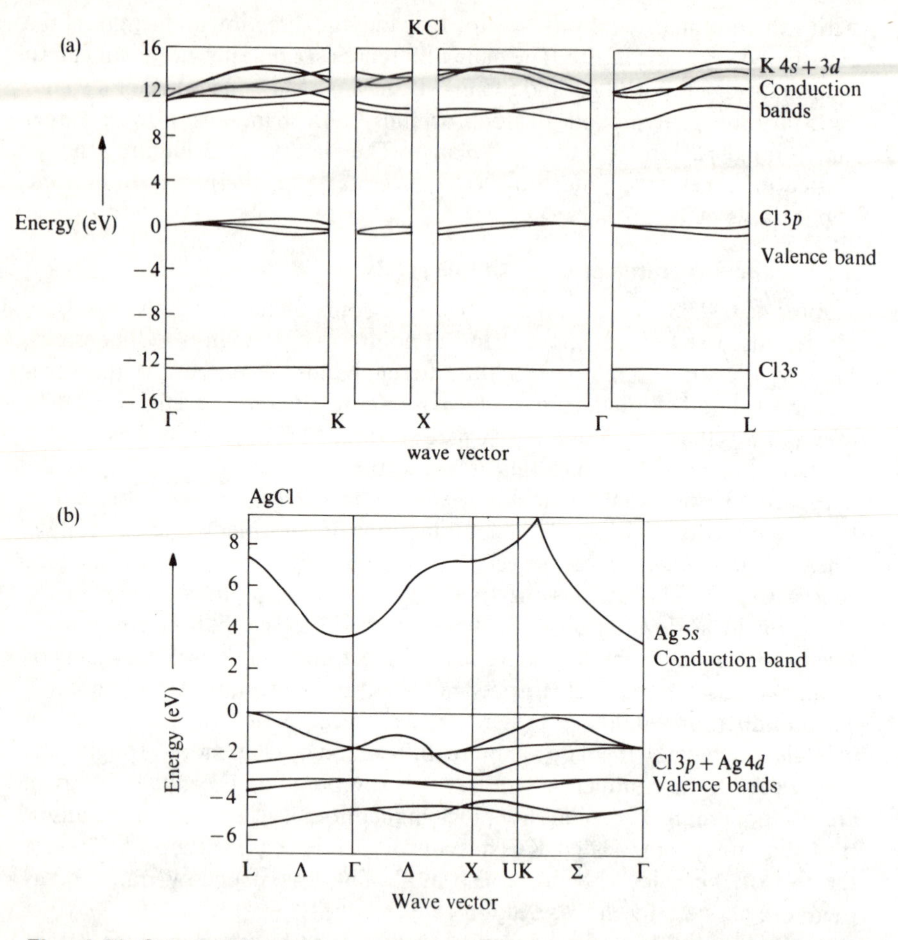

Fig. 4.30 Calculated band structures for KCl and AgCl. The labels on the k axis show different directions for the wave-vector. In KCl the K *3d* levels are empty and mix with the conduction band; in AgCl the filled *4d* levels mix with the Cl *3p* valence band orbitals. The indirect band gap of AgCl can be seen. (From M. L. Cohen and V. Heine, *Solid State Phys.*, 24 (1970), 169; J. Shy-Yih Wong, M. Schluster, and M. L. Cohen, *Phys. Stat. Solidi* (*B*), **77** (1976), 295.)

valence band energy occurs for different **k** values. Silver chloride therefore has an indirect band gap, and the lowest energy transition is forbidden by the k selection rule. The indirect band gaps in silver halides inhibit the rapid recombination of electrons and holes. This seems to be important in their applications in photography, discussed briefly in Chapter 7.

Another solid showing covalent mixing of p and d orbitals is the transition-metal compound rhenium trioxide, in Fig. 4.31. The rhenium $5d$ orbitals now

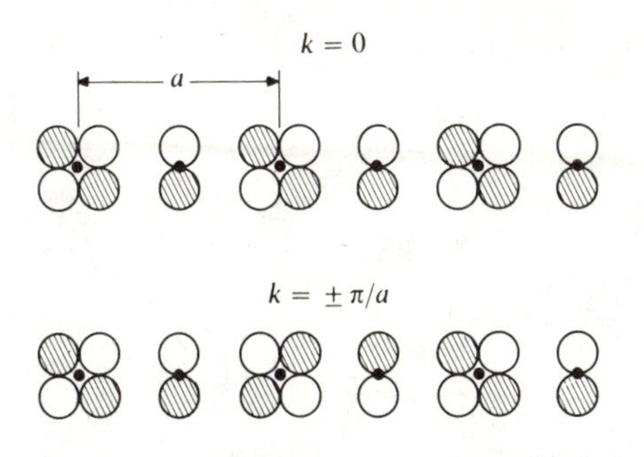

Fig. 4.31 At $k = 0$ there is no net interaction between *p* and *d* orbitals on different atoms; maximum interaction occurs at $k = \pm\pi/a$.

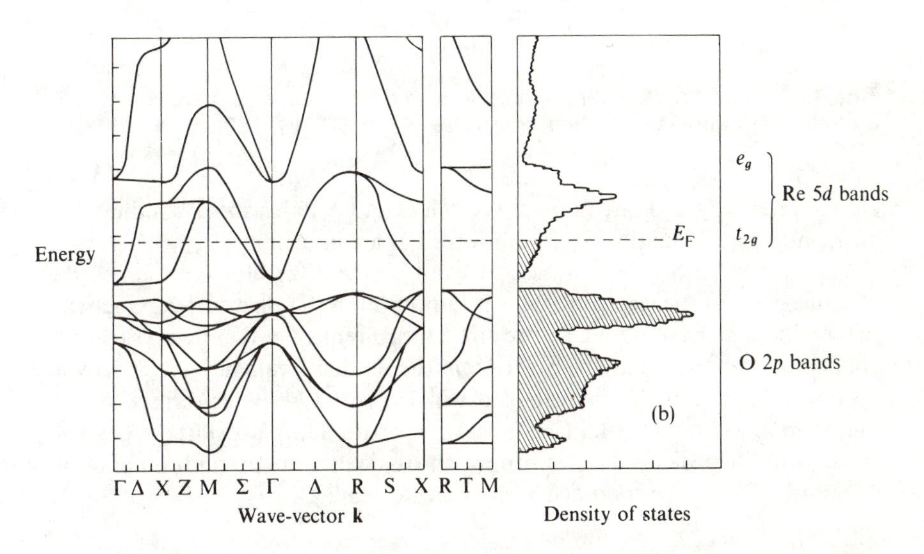

Fig. 4.32 Calculated band structure for ReO_3, with the density of states and the Fermi level in the Re 5*d* band. (From L. F. Matthies, *Phys. Rev.*, **181** (1969), 987.)

form the conduction band, and are partially occupied, making ReO_3 metallic. The valence bands are predominantly oxygen 2*p* orbitals: the rather complicated appearance of these bands in the diagram arises because there are three oxygen atoms in the unit cell, making altogether nine atomic orbitals. The conduction band is really an antibonding combination of the oxygen 2*p* with the metal 5*d* orbitals. Just as with AgCl, no mixing is possible at Γ, where $k = 0$. So at this wave-vector, the conduction band orbitals are in fact pure Re 5*d* combinations, and the valence band is purely O 2*p*. This is why the bands are

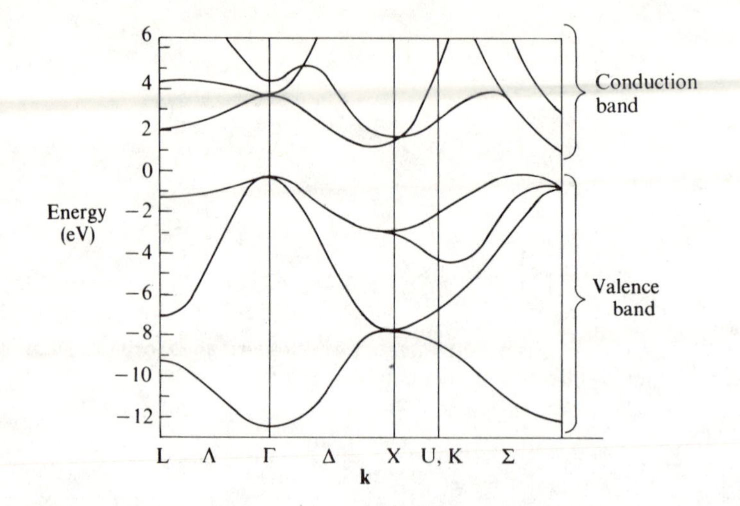

Fig. 4.33 Band structure for silicon, showing the indirect band gap. (From J. R. Chelikovsky and M. L. Cohen, *Phys. Rev. B*, **14** (1976), 100.)

closest in energy at Γ: at non-zero **k** values the orbitals form bonding and antibonding combinations, and are forced apart in energy.

The curvature of the conduction band in the $E(\mathbf{k})$ plot determines the effective mass of the electrons. This is important in the electrical conductivity of the solid, but in fact a much more direct measurement of the effective mass of conduction electrons may be made from the plasma frequency. As we saw in Section 2.4.3, this is the frequency at which a metal becomes transparent to light, and is thus important in its optical properties. The plasma frequency also corresponds to peaks in the electron energy loss spectrum. In solid compounds, equation 2.7 must be modified somewhat, to read:

$$\omega_p = (Ne^2/\varepsilon_0\varepsilon_{opt}m^*)^{1/2}.$$

N is the concentration of metallic electrons, and ε_{opt} the high-frequency dielectric constant, determined by the polarizability of electrons in the valence band. If these quantities are known, the effective mass m^* may be deduced. This gives a useful comparison with calculated band structures, and at a simpler level, provides a qualitative measure of the width of the conduction band. In some metallic oxides such as ReO_3, the covalent bonding is strong enough to give rather broad d bands, with effective masses similar to the free electron mass.

Figure 4.33 shows the band structure for silicon. In Section 3.2.1 of Chapter 3, the electronic structure of this tetrahedral solid was described in terms of the overlap of 3s and 3p valence orbitals. The band structure gives a more detailed picture. At zero wave-vector, s and p orbitals cannot mix, because of their different symmetry. (This is similar to the case of p and d orbitals discussed above). Thus the valence bands at Γ in Fig. 4.33 are pure s (lowest in energy) and p combinations. Away from Γ, mixing is allowed, and the orbital combinations are now bonding mixtures of both s and p orbitals. It can be seen that silicon also has an indirect band gap. As with the silver halides, this fact is important in some of its applications.

4.3.2 Metals and alloys

For metallic solids the chemist's picture of localized bonding electrons is not applicable, and the band structure approach becomes more essential for describing the electronic structure. In this section, we shall describe some applications of band theory to metals and alloys.

Figure 4.34 shows some features of the electronic structure of copper, which has the cubic close-packed (f.c.c.) structure. The band structure diagram confirms the essential features discussed in the Chapter 3. There is a full set of rather narrow 3*d* bands, overlapped in energy by a broad band, coming from the overlap of the more diffuse 4*s* and 4*p* orbitals. The dashed curve in Fig. 4.34(a) shows the behaviour expected for the *s*–*p* band if the 3*d* orbitals were

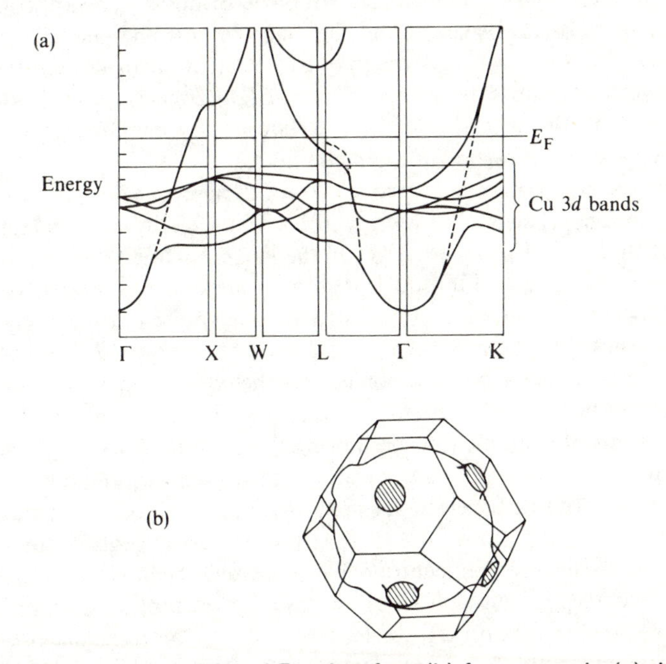

Fig. 4.34 Band structure (a) and Fermi surface (b) for copper. In (a) the Cu 3*d* bands are labelled; the dashed curve shows the *4s* band predicted without any mixing with the *d* band. (From G. A. Burdick, *Phys. Rev.*, **129** (1963), 138.)

ng, and the $E(\mathbf{k})$ curves are similar to those for nearly free electrons in a periodic potential. The full curves include the effect of mixing between the s–p band and the d levels, when their energies are similar.

In the two-dimensional models discussed earlier in this chapter, we showed some energy contours in the (k_x, k_y) plane. The analogue of the contour in three dimensions would be a constant energy *surface*, shown as a function of $\mathbf{k}$. It is not easy to show several successive energy surfaces on the same diagram, and this method of showing how E varies with $\mathbf{k}$ is not commonly used in three dimensions. There is one constant energy surface, however, which is very important. This is the **Fermi surface** of a metal, which shows the wave-vectors for electrons with energies just at the Fermi level. For free electrons, the energy is given by a simple generalization of equation 4.70:

$$E(\mathbf{k}) = V_0 + (\hbar^2/2m)(k_x^2 + k_y^2 + k_z^2)$$

Since the energy depends only on the magnitude of $\mathbf{k}$ for free electrons, any constant energy surface, including the Fermi surface, is a sphere. We saw in two dimensions however, that the periodic lattice potential distorts the $E(\mathbf{k})$ curves near values where the electron waves just fit into the lattice. This happens when $\mathbf{k}$ is at a Brillouin zone boundary. Figure 4.34(b) shows the Fermi surface for copper. The truncated octahedron shown is the Brillouin zone appropriate to the f.c.c. lattice. It can be seen from the diagram how the periodic potential distorts the $E(\mathbf{k})$ function from the pure free-electron form, so that the Fermi surface is pulled towards the hexagonal faces of the zone. It is possible to verify this feature of the band structure of copper by measurements of the conductivity in the presence of magnetic fields.

A weak periodic potential in two and three dimensions does not produce a band gap, but nevertheless, as we saw in Section 4.2.2, it has an effect on the density of states (see Fig. 4.23(b)). When the Fermi surface touches a Brillouin zone face, $E(\mathbf{k})$ flattens out, and this gives a peak in the density of states, followed by a drop. Different crystal structures give rise to differently shaped Brillouin zones, and the number of electrons required to touch the zone face will vary. Thus in nearly-free electron metals, the positions of peaks and dips in the density of states should depend in a fairly simple way on the structure. This forms the basis of a simple interpretation of the **Hume-Rothery rules**, which show how the crystal structures of alloys can often be predicted from the average number of valence electrons per atom. For example, CuZn and Cu_3Al, each with an electron/atom ratio of 3/2, have the body-centred cubic (b.c.c.) structure. As the electron concentration is increased, the more complex γ-brass structure becomes favoured. This is the case, for example, with Cu_5Zn_8 and Cu_9Al_4, each with a ratio 21/13. Higher still, the hexagonal close-packed arrangement is found, for example in $CuZn_3$ and Cu_3Si, where the electron/atom ratio is 7/4.

Consider for two different crystal structures, for example f.c.c. and b.c.c. In the f.c.c. structure it can be shown that the free-electron Fermi surface touches the Brillouin zone edge when there are 1.36 electrons per atom, but that for the b.c.c. structure, 1.48 electrons are required. Approximate densities of states are shown in Fig. 4.35. For a solid with up to about 1.4 electrons per atom, the f.c.c. structure will be favoured, but because of the drop in density of states at this point, any more electrons must be accommodated in rapidly increasing energy levels in the band. Thus the average electron energy will now be greater than in the b.c.c. structure. B.c.c. will be satisfactory up to an electron/atom ratio of about 1.5, after which this structure in turn becomes less stable. In the γ-brass structure, the number of electrons required to touch the zone boundary is predicted to be 1.54, which explains why this is adopted by alloys with an electron/atom ratio of around $21/13 = 1.6$, but not at higher concentrations.

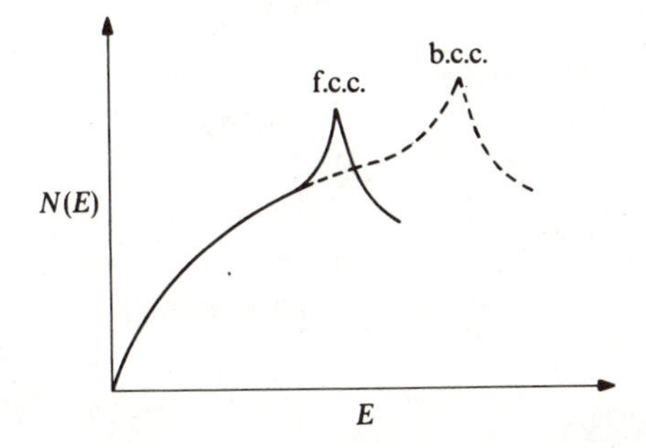

Fig. 4.35 Schematic diagram of nearly-free electron density of states curves for f.c.c. and b.c.c. lattices. The peak followed by a dip occurs when the free-electron wave-vector touches the Brillouin zone boundary.

The interpretation of the Hume–Rothery rules in terms of free-electron theory is very appealing, but unfortunately it cannot be quite correct as it stands. Copper has the f.c.c. structure, with one valence electron per atom. In the free-electron theory, the Fermi surface should be well inside the Brillouin zone boundary. However, Fig. 4.34 showed that the Fermi surface is in fact distorted so much that it touches the boundary. Thus the effect of the periodic potential is much more serious than the simple theory assumes. The actual density of states curves are rather more complicated than in the naïve picture of Fig. 4.35. Another unsatisfactory feature is that explanation neglects the effect that atoms of different nuclear charge must have on the band structure. The basic idea that it is the shape of the Brillouin zone in the different structures that is important, may well be correct. A fully satisfactory explanation of the Hume–Rothery rules, however, must be a little more sophisticated than the one discussed here.

Calculated density of states curves including the d levels have been used in a rather similar way to explain the changes in metallic structure observed across the transition series. The early transition metals, Sc, Ti and elements below them, have h.c.p. structures. These are followed by b.c.c. solids (the vanadium and chromium groups), and later by hexagonal close-packed (h.c.p.) and f.c.c. metals. (Manganese has a complex structure, and iron falls out of line, with the b.c.c. structure being most stable at room temperature. However, this seems to be associated with its ferromagnetism, as it reverts to the f.c.c. structure above the Curie temperature.) Figure 4.36 shows calculated densities of states for a model transition metal, comparing the b.c.c., f.c.c. and h.c.p. structures. The shapes of these curves reflect the different overlaps possible between d orbitals in the different structures. The broad peak in the b.c.c. density of states occurs when there are up to six valence electrons per atom, and means that the average electron energy will be lower for the b.c.c. structure at this electron count. However, the peak is succeeded by a deep minimum, and at this point, one of

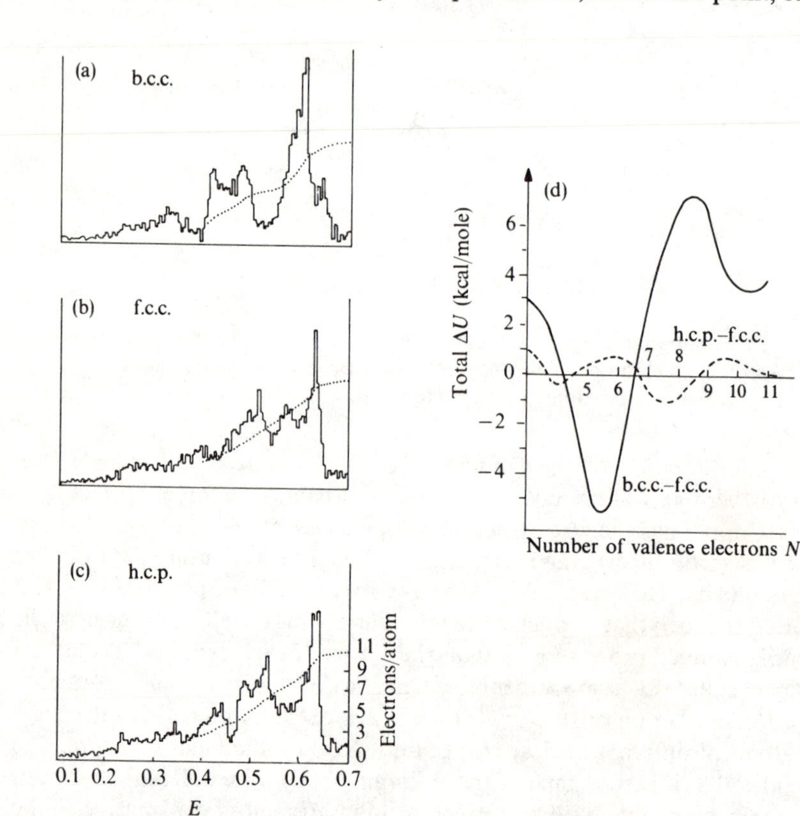

Fig. 4.36 Calculated density of states curves for model transition-metal elements in the (a) b.c.c., (b) f.c.c., and (c) h.c.p. structures. (d) Predicted relative energy of different structures, as a function of the number of valence $(s+d)$ electrons. (From D. G. Pettifor, *Calphad,* **1**, (1977), 305.)

the close-packed structures becomes more stable. This is shown in Figu 4.36(d), where the difference in the total electronic energy, predicted from th density of states for the different structures, is plotted against the number o valence electrons. The largest differences of energy are between the b.c.c. and the close-packed structures. For electron counts of five and six, the electronic energy given by the b.c.c. structure is lowest, which is in agreement with the observation of this structure in the vanadium and chromium groups. For seven or more electrons, there is a close balance between f.c.c. and h.c.p. structures. However, the h.c.p. structure is correctly predicted to be stable for metals with seven and eight valence electrons, and the f.c.c. for later transition elements. These calculations show how the structures of metallic elements are determined by rather subtle differences in the density of states, which in turn are controlled by the different types of bonding interaction present. Attempts have been made to give a more 'pictorial' account of some of these factors. For example, the formation of the b.c.c. structure by elements in groups V and VI has been discussed using a model of bonding with *s–d* hybrids. Unfortunately, the calculations just discussed do not support this theory, as the energy differences depend almost entirely on bonding by the *d* electrons, and interaction with the *s* band has only a minor influence on the predicted structure.

4.3.3 Chemical interpretations: orbital and overlap populations

Although we have tried, in the preceding sections, to show how band-structure diagrams can be interpreted in chemical terms, it has to be admitted that the 'spaghetti-like' curves such as those of Fig. 4.32 are rather unappealing to the chemist. Even the total density of states curves calculated from band structures do not immediately give chemical bonding information. It is interesting to see, therefore, how the results of band-structure calculations can be presented in a chemically more informative way.

Let us return for a moment to the one-dimensional model of a diatomic solid, the binary chain presented in Section 4.1.3 (p. 88): We showed how the crystal orbitals were constructed as linear combinations of atomic orbitals from the two kinds of atom in the chain (equation 4.30):

$$\psi_k = \sum_n \exp(inka)[a_k \chi(A)_n + b_k \chi(B)_n].$$

The coefficients a_k and b_k give the amplitudes, respectively, of the A and B atomic orbitals in ψ_k. In fact, in the simple Hückel theory where overlap integrals are neglected, the *squares* of these coefficients are equal to the electron *density* on the atoms, for an electron in this crystal orbital. The total atomic populations could be found by summing these coefficients over all occupied orbitals. However, it is more revealing to show the atomic contributions to different orbitals in the bands, by weighting the density of states curve by the appropriate values of $(a_k)^2$ and $(b_k)^2$. This is done in Fig. 4.37. The top curve (a) shows the total density of states coming from the two bands of Fig. 4.8 on p. 90. The next two plots, (b) and (c), show the atomic populations. As explained in Section 4.1.3, the lower band has more contribution from the electronegative atom B, and the upper band more from the electropositive A. Figure 4.37 shows how the atomic populations vary through the bands. The covalent interaction between the two kinds of atom is shown by the occurrence of some B population in the upper band, and some A population in the lower one. Such covalency can also be illustrated by looking at the product of the coefficients: $(a_k)\cdot(b_k)$. For bonding overlaps, we expect neighbouring atomic orbitals be combined with the same sign, so that the product is positive. Antibonding orbitals would correspondingly give a negative value. Figure 4.37(d) shows the density of states weighted by this overlap product. The diagram shows how the band at lower energy is composed of A–B bonding orbitals (+), and the higher energy band is made up of antibonding ones (−).

The idea just illustrated has been used to present the results of band structure calculations on a number of compounds where the bonding is

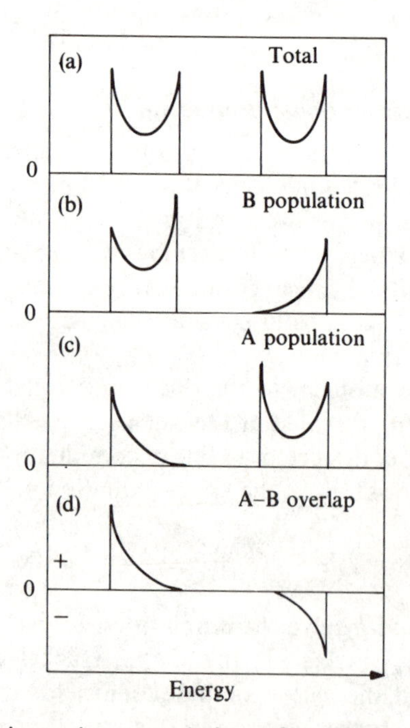

Fig. 4.37 Orbital and overlap populations for bands formed by the one-dimensional binary chain (see Fig. 4.8). (a) Total density of states; (b) and (c) orbital populations on B and A atoms respectively; (d) A–B overlap population, showing bonding (+) and antibonding (−) regions.

relatively complex. Many of these calculations have been performed by the **extended Hückel method**, which is similar to the simple model used earlier in this chapter, in that parameterized values are taken for the Hamiltonian matrix elements between atomic orbitals (see equation 4.20). These are generally estimated from atomic ionization potential data. The important difference in the extended Hückel method is that overlap integrals between atomic orbitals are calculated explicitly, and included in the calculation of crystal orbital energies. The product of orbital coefficients (a_k) (b_k) in these calculations is multiplied by the overlap integral between A and B orbitals, and the result called the **crystal orbital overlap population** (COOP). An example of this kind of plot is shown in Fig. 4.38. The compound NbO has a structure based on that

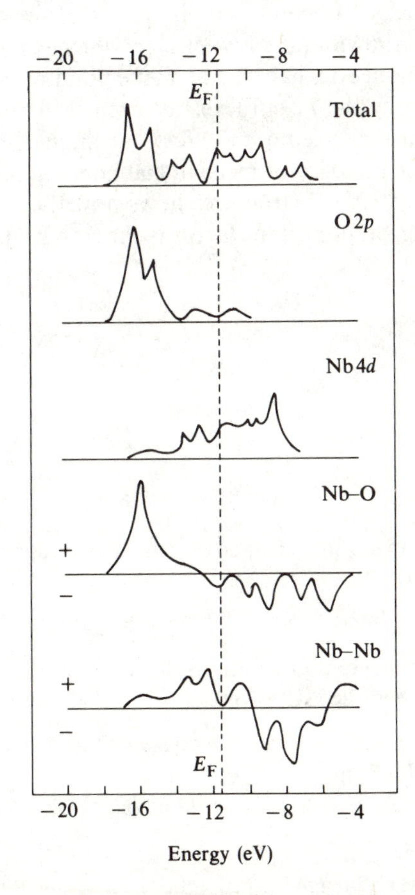

Fig. 4.38 Orbital and overlap populations from a band-structure calculation on NbO. The Fermi level corresponding to the $4d^3$ electron configuration is shown. (From J. K. Burdett and T. Hughbanks, *J. Am. Chem. Soc.*, **106** (1984), 3101.)

of NaCl, but with one in four lattice sites vacant, to form an ordered defect structure. From an ionic point of view, such a vacancy structure is very unfavourable, because of the loss of Coulomb bonding energy. It has generally been assumed that significant metal–metal bonding between niobium atoms is present, and helps to stabilize the unusual structure. The top three plots of Fig. 4.38 show the total calculated density of states of valence electrons, and its breakdown into O 2*p* and Nb 4*d* contributions. This shows how the predominantly O 2*p* band at around 16 eV binding energy is succeeded by a band of mostly Nb 4*d* character at higher energy. The Fermi level shows the occupancy of orbitals in the d^3 compound. The COOP plots at the bottom of the diagram show more clearly the bonding character of the different levels. The 'O 2*p* band', as expected, has significant niobium–oxygen bonding properties, whilst the 'Nb 4*d* band' is antibonding. However, it is the Nb–Nb bonding interaction, shown in the lowest plot, which is most interesting. The lower part of the 4*d* band does have metal–metal bonding character, and the d^3 configuration shown by the Fermi level represents just the right number of electrons to obtain the maximum effect, as higher levels are Nb–Nb antibonding. Thus the presence of metal–metal bonding is confirmed. A comparison with the simple NaCl structure shows how the vacancies are able to enhance this bonding; an important factor is the stabilization of some Nb 4*d* atomic orbitals.

Further reading

Most accounts of band theory are based on the free-electron approach. An elementary, tutorial, account is given in:

S. L. Altmann (1970). *Band theory of metals*. Pergamon Press.

The theory of Brillouin zones is described in the following (the book by Kittel also treats the conductivity and the Hall effect):

H. Jones (1975). *The theory of Brillouin zones and electronic states in crystals* (2nd edn). North-Holland.

C. Kittel (1976): *Introduction to solid state physics* (5th edn), Chapter 2. John Wiley and Sons.

An account of electrical and magnetic properties is given by:

B. I. Bleaney and B. Bleaney (1976). *Electricity and magnetism* (3rd edn) Chapters 11, 12. Oxford University Press.

A classic account of the theory of metals, containing a discussion of the Hume–Rothery rules, is:

N. F. Mott and H. Jones (1936). *The theory of the properties of metals and alloys.* [Oxford University Press; reprinted by Dover Publications.]

Applications of an LCAO approach similar to that described in the present chapter are given in the articles:

J. K. Burdett (1984). *Prog. Solid State Chem.* **15** 173.
M. Kertesz (1985). *Int. Rev. Phys. Chem.* **4** 125.

The following papers describe applications of the extended Hückel method and the COOP plot to band-structure calculations:

R. Hoffmann *et al.* (1978). *J. Am. Chem. Soc.* **100** (1983). *J. Am. Chem. Soc.*, **105** 1150–62, 3528–37.
J. K. Burdett and T. Hughbanks (1984). *J. Am. Chem. Soc.* **106** 3101.

The measurement of band structures by photoelectron spectroscopy is described, with many examples, in:

N. V. Smith (1978). In *Photoemission in solids*: *I* (ed. M. Cardona and L. Ley). *Topics in applied physics*, Vol. **26**. Springer Verlag.
L. Ley, M. Cardona, and R. A. Pollak (1979). In *Photoemission in Solids*: II (ed. L. Ley and M. Cardona), *Topics in applied physics*, Vol. **27**. Springer Verlag.

5 The effects of electron repulsion

In the orbital pictures of atoms and molecules, electrostatic repulsion between electrons is treated in an approximate way. It is assumed that electrons move independently, in a potential field that includes the average repulsion from the other electrons. The same approximation is made in band theory. Although electron repulsion was not explicitly mentioned in Chapter 4, it is implicitly included, either in the atomic orbital energies which enter the LCAO calculations, or in the periodic potential for the nearly-free electron model. The orbital approximation works very well in many situations, but there are many where it cannot be used. For example, molecular orbital theory can break down rather seriously if the interaction between atomic orbitals is weak, such as in the dissociation of a molecule. The same problem arises in solids, whenever the inter-atomic overlap is small, so that only rather narrow bands are formed. The effects of electron repulsion then become much more important, and cannot be treated simply as an average potential. Many compounds of transition elements and lanthanides are not metallic, although they appear to have partially filled *d* or *f* bands. In these compounds, band theory does not work, and the electrons are localized by their mutual repulsion. Since the class of compounds where this happens is a very large and important one, we shall discuss in some detail here how electrons become localized, and how this affects their electronic properties.

5.1 The Hubbard model

The full treatment of electron repulsion effects in molecules and solids is extremely difficult, and fairly gross approximations have to be made in any but the simplest situations. The most fruitful approach to electron repulsion in solids is the **Hubbard model**, which assumes that the only important repulsion effects occur between two electrons on the *same* atom. Although the repulsion between electrons on different atoms is certainly not negligible, the intra-atomic effects seem to be chiefly responsible for the breakdown of band theory. The Hubbard model therefore provides a very useful picture for discussing electron localization.

Imagine an array of atoms, each with a single valence *s* orbital. In Chapter 4, it was shown how the bonding interaction gives rise to a band of crystal orbitals, each delocalized throughout the solid. If there is one electron per atom, the band is half full, and band theory predicts that the solid should be metallic. But consider the situation where the overlap between orbitals is very small. The ground state will now have one electron *localized* on each atom. This is because of the extra electron repulsion that results when we try to move an electron to another atom, as shown for a one-dimensional array in Fig. 5.1. The energy required to remove an electron from one atomic orbital is I, the ionization energy. In placing the electron on a site already occupied, we get back an energy A, the electron affinity of the neutral atom. The energy required to move the electron is therefore given by:

$$U = I - A. \tag{5.1}$$

This quantity can be interpreted as the repulsion energy between two electrons on the same atom. For hydrogen, $I = 13.6$ eV and $A = 0.8$ eV, which gives $U = 12.8$ eV.

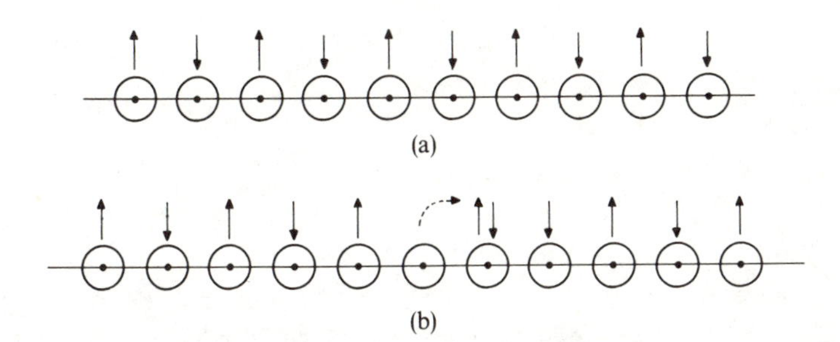

Fig. 5.1 The effects of electron repulsion in a chain of weakly interacting atoms. (a) Ground state with one electron localized in each atomic orbital. (b) Moving an electron involves extra electron repulsion because of the doubly-occupied orbital.

The effect of electron repulsion is to make the half-filled band *insulating*, when the interaction between atoms is small. The situation as the bandwidth increases is illustrated in Fig. 5.2(a). On the left, where the bandwidth W is zero, are the atomic energies. The lower energy ($-I$), is the energy of the singly occupied orbitals. The upper energy ($-A$ in the diagram), is the energy of an extra electron added to the solid, to give a doubly occupied orbital. The gap, equal to U in equation 5.1, shows the energy required to excite an electron, so that it can move to another orbital. This gap is *not* the same as that which occurs in band theory, but is a consequence of electron repulsion. The lower and upper levels, each holding *one* electron per atom, could be called **sub-bands**. The gap U is often called the **Mott–Hubbard** splitting, after the two

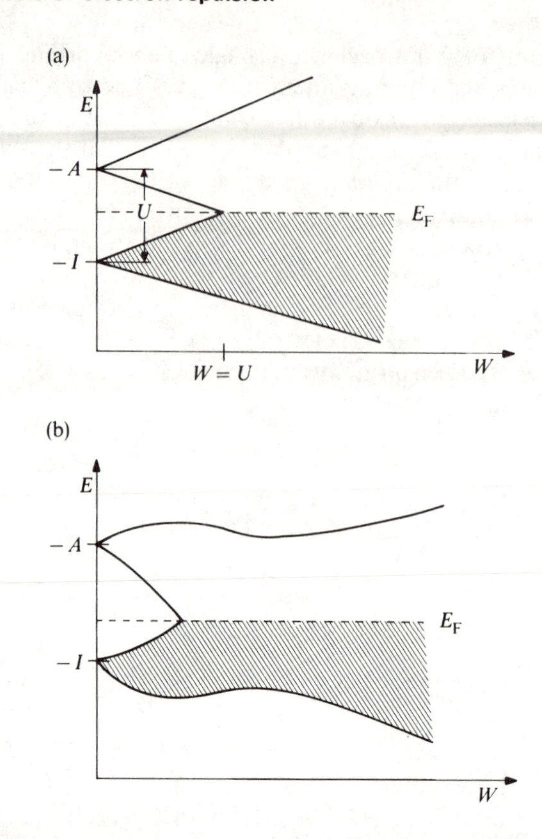

Fig. 5.2 Hubbard sub-bands, as a function of bandwidth, W. At $W = 0$ the energies are just those of adding an electron (electron affinity A) and removing an electron (ionization energy I). (a) The overlap of sub-bands when $W > U$. (b) More realistic plot, including the decrease of U due to increased polarizability of the solid as W increases.

physicists who have contributed most to the theory of electron localization in solids. It can be seen in Fig. 5.2(a) that each sub-band is broadened by the inter-atomic overlap, and that they meet when the width W and the repulsion parameter U are approximately equal. Beyond this point, there is no energy gap, and the solid is metallic. Thus for band theory to work, we require that:

$$W > U \tag{5.2}$$

The parameter U therefore plays a fundamental role in the Hubbard model. For isolated atoms, it is given by an equation such as (5.1). As discussed earlier, however, there are important polarization effects in solids, whenever an electron is moved from one site to another. The polarization of other electrons, towards the vacant hole, and away from the extra electron, considerably lowers the energy required to excite an electron. Values of the 'Hubbard U' measured in solids are much smaller than the atomic energies would suggest. As the orbitals overlap more, the polarizability of the solid increases, since electron clouds can distort, not only on one site, but also by moving partially between atoms. As the bandwidth increases, the value of U therefore tends to decrease. A more realistic picture of the Hubbard sub-bands, including this change of U, is shown in Fig. 5.2(b).

There is an interesting connection between the Hubbard model and a simple treatment of diatomic molecules that also takes account of electron repulsion. For H_2, the bonding molecular orbitals are normally considered to be symmetrical combinations of the 1s atomic orbitals on the two atoms A and B:

$$\psi = \chi_A + \chi_B. \tag{5.3}$$

The molecular orbital wave function is:

$$\Psi(1,2) = \psi(1)\psi(2), \tag{5.4}$$

and describes the two electrons delocalized equally between the atoms. However, as the molecule dissociates, this wave function becomes a very bad approximation, and repulsion between electrons causes them to become localized on each atom, just as in a solid. A simple modification to the MO theory was made by Coulson and Fischer, who wrote the wave function in terms of *unsymmetrical* combinations of atomic orbitals:

$$\begin{aligned} \psi(1) &= \chi_A + \mu\chi_B \\ \psi(2) &= \chi_B + \mu\chi_A \end{aligned} \tag{5.5}$$

The parameter μ in the Coulson–Fischer wave function is variable, and describes the degree of localization of each electron. When $\mu = 1$, we have the normal MO wave function, which is appropriate to strong overlap between the orbitals. As the overlap interaction diminishes, electron repulsion becomes more important, and μ decreases. In the fully dissociated molecule, μ is zero, corresponding to complete localization. This is closely analogous to the prediction of the Hubbard model. The MO wave function of a diatomic corresponds to the Bloch functions of band theory. As the orbital overlap diminishes, the repulsion effect progressively localizes electrons on individual atoms.

5.2 Lanthanides

By far the clearest illustration of the Hubbard model is provided by solids containing elements from the lanthanide series, cerium to lutetium. We shall therefore look at this series before considering transition-metal compounds, where bandwidth and electron repulsion effects are often in competition, so that the behaviour can be quite complicated. Across the lanthanide series, the

4*f* shell is progressively filled. In most of the elements, the 4*f* orbitals are highly contracted, and have only a very small overlap with other valence orbitals. This may not be quite true near the beginning of the series, and recent measurements of the electronic structure of cerium and its compounds suggest some degree of interaction between the Ce 4*f* orbital and its surroundings. For the later lanthanides, however, there is no doubt that the 4*f* levels are highly localized. It is this which gives rise to the remarkable chemical similarity of the elements, since the partially filled 4*f* shell makes very little contribution to the chemical bonding. The filling of the 4*f* orbitals does have an influence, however, on the oxidation states available. The most stable ions in aqueous solution are the Ln^{3+}, but some of the elements also show 2+ or 4+ states. The trend in 4*f* energies responsible for this also affects the electronic structure of the solid elements and their compounds.

The complete localization of the 4*f* electrons is shown by many physical properties of lanthanide compounds. For example, the magnetic susceptibilities are very close to the values predicted for the free ions. Spectroscopic measurements show the states predicted for configurations of 4*f* electrons, with very little influence of the surrounding atoms. Small ligand-field effects can be seen in the spectra, but the splittings (around 100 cm^{-1} or 0.01 eV) are some two orders of magnitude smaller than with the transition elements. Most solid compounds are identical in these respects to complexes observed in dilute solution. All the evidence suggests that, except possibly in cerium at the beginning of the series, the 4*f* bandwidths are negligible. It is primarily for this reason that the lanthanide series gives such a clear illustration of the applications of the Hubbard model.

5.2.1 4*f* energies and the Hubbard *U*

The *f* shell can hold a maximum of 14 electrons, and in most of the lanthanides is partially occupied. Thus the electron transfer process defining the *U* parameter is now a little more complicated than in *s* orbitals considered in the previous section. In a majority of solids, the lanthanide has a configuration $4f^n$ corresponding to the trivalent ion Ln^{3+}. The energy required to move an electron corresponds to the process:

$$2Ln^{3+} \rightarrow Ln^{2+} + Ln^{4+}, \tag{5.6}$$

or

$$2(4f)^n \rightarrow (4f)^{n+1} + (4f)^{n-1}. \tag{5.7}$$

For free ions, the energy of this process is given by the difference between the fourth and third ionization potentials:

$$U = I_4 - I_3. \tag{5.8}$$

(Actually, in two cases, lanthanum and gadolinium, the Ln^{2+} ion has the ground-state configuration $(4f)^n(5d)^1$. The $(4f)^{n+1}$ configuration is quite close, however, and a small correction can be made to equation 5.8.) For gas-phase ions, predicted values of U are around 25 eV. In solids, however, we expect them to be reduced by polarization effects.

The $4f$ orbitals appear strongly in photoelectron and inverse photoelectron spectra measured with photons in the X-ray region. Some of the spectra are quite complicated, because of the spin and orbital coupling between different $4f$ electrons in the same atom. However, the techniques give a direct measurement of the change in $4f$ orbital energies across the series. Results have been obtained for some compounds, but the most clearly resolved spectra come from the metallic elements themselves. Even in the metals, the $4f$ orbitals are localized, and do not contribute to the bonding. The metallic nature comes from overlap of the more diffuse $5d$ and $6s$ orbitals, which are close in energy to the $4f$. Thus the photoelectron spectra of the $4f$ levels can be interpreted in terms of highly localized, atomic-like, levels.

Figure 5.3 shows a combined picture of the photoelectron and inverse photoelectron spectra of gadolinium. The $4f$ bands are clearly visible in each spectrum. In PES, an electron is ionized, and the measured binding energy corresponds to the process:

$$(4f)^n \rightarrow (4f)^{n-1}. \tag{5.9}$$

In the inverse spectrum, as explained in Section 2.2.3, an electron is added to the solid, and the peak corresponds to the process:

$$(4f)^n \rightarrow (4f)^{n+1}. \tag{5.10}$$

The two spectra are lined up at the Fermi level, where the photoelectron spectrum ends, and the inverse spectrum which shows empty levels begins. Thus the separation between the $4f$ peaks in the spectra is just the energy required to remove an electron from one $4f$ orbital, and to put it into another one: that is, the energy U defined in equation 5.8. The value found for

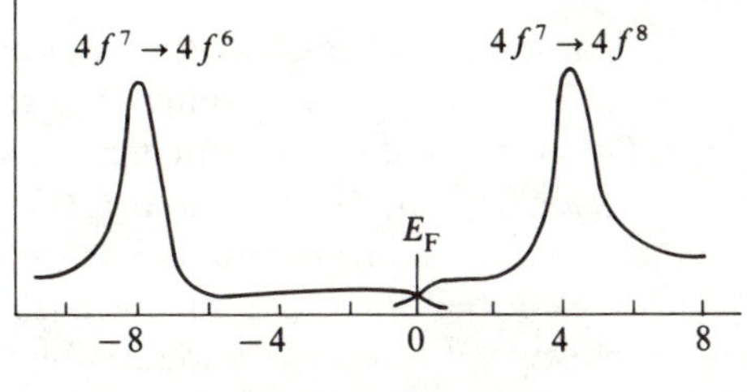

Fig. 5.3 Photoelectron ($E < E_F$) and inverse photoelectron ($E > E_F$) spectra of gadolinium metal. The $4f$ configurations measured in each technique are shown.

gadolinium is 12 eV, which is *12 eV* less than the gas-phase value. This illustrates clearly the importance of solid-state polarization effects.

Figure 5.4 shows a complete plot of the Hubbard U values for the $4f$ orbitals in lanthanide metals (except for promethium, which has no stable isotope). All the values are greater than 4 eV. Although the $4f$ bandwidths are difficult to measure, they are almost certainly less than 0.1 eV, and it can easily be understood why the electrons are so localized. Values of U have been obtained for some solid compounds, and are very similar to those in the elements.

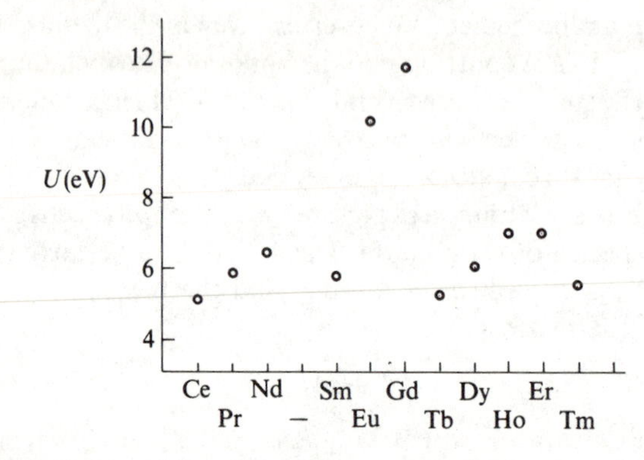

Fig. 5.4 Hubbard *U* values measured for the lanthanide metals.

The values plotted in Fig. 5.4 show an interesting pattern, which can be explained by looking at the measured energies for the individual $(4f)^n$ to $(4f)^{n-1}$ and $(4f)^{n+1}$ processes, plotted in Fig. 5.5. The irregular trends reflect very closely those found in the gas-phase ionization energies, and are a result of the differences in repulsion between $4f$ electrons as the shell is successively filled. There are three contributions:

(i) The increasing nuclear charge across the series gives a steady stabilization (trend to higher binding energy, and higher electron affinity).

(ii) There is a break at the half-filled shell. Up to seven electrons can be placed in different orbitals with parallel spins, thus giving a favourable exchange energy. The eighth and subsequent electrons lose this stabilization, as they must pair up with electrons in orbitals already occupied.

(iii) The effects of angular momentum coupling between different $4f$ electrons give a pronounced curvature to the plots. This is also an electron repulsion effect, and is more pronounced in the second half of the series, where the orbitals are more contracted.

The $(4f)^n$ configuration referred to in the plot in Fig. 5.5 is that of the Ln^{3+} ion. Most of the metals have this configuration, together with three metallic

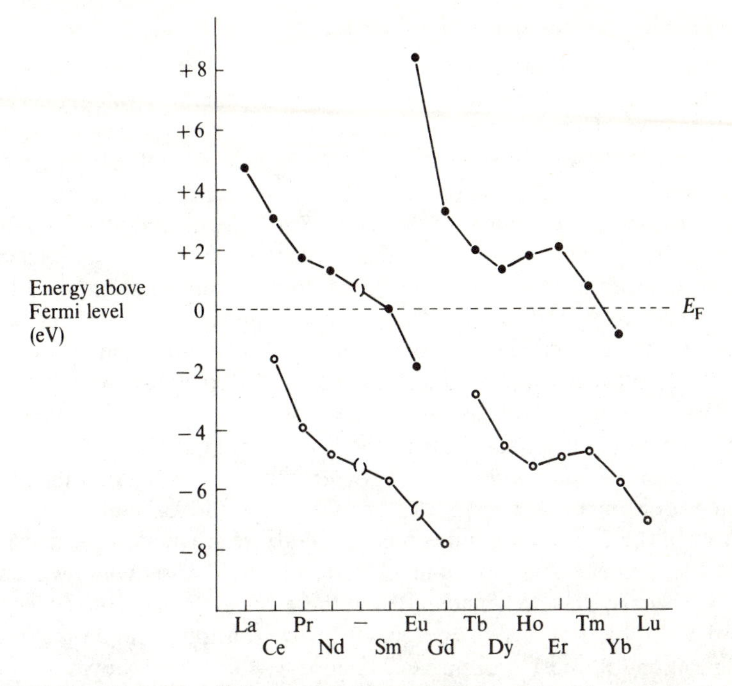

Fig. 5.5 Energies of the $(4f)^n$ to $(4f)^{n+1}$ and $(4f)^n$ to $(4f)^{n-1}$ processes relative to the Fermi level in lanthanide metals. In each case $(4f)^n$ is the configuration of the trivalent ion Ln^{3+}. ●, $4f^n \rightarrow 4f^{n+1}$; ○, $4f^n \rightarrow 4f^{n-1}$.

electrons in the broad (*s–d*) band. However, in europium and ytterbium, the energy of the $(4f)^n$ to $(4f)^{n+1}$ process has fallen below the Fermi level. The $4f$ shell in these elements therefore captures another electron, leaving only two to contribute to the metallic bonding. In fact metallic Eu and Yb have lower sublimation energies, and larger atomic volumes, than the other lanthanides in the series. The presence of the $4f^7$ (Eu^{2+}) and $4f^{14}$ (Yb^{2+}) configurations is confirmed by magnetic measurements.

The plot of U values in Fig. 5.4 shows that exceptionally high values are found for Gd and Eu, where the ground state is the half-filled shell $4f^7$. In these elements an electron is removed from a configuration with a maximum number of parallel spins, and so with a maximum exchange stabilization. This stabilization is totally lost when it is placed on another ion, because it must now have a spin opposite to that of all the other electrons present. A similar loss of exchange stabilization does not happen for other electron configurations. The extra contribution to U for ions with a half-filled shell is also important in the transition series.

5.2.2 Divalent compounds of lanthanides

The trends in $4f$ binding energies shown in Fig. 5.5 can be used to discuss some interesting features of lanthanide chemistry. Although the trivalent Ln^{3+} state is the commonest, other oxidation states are observed. At the beginning of the series (Ce and Pr) and after the break at the half-filled shell (Tb), the $(4f)^n$ to $(4f)^{n-1}$ ionization energy is low enough for some Ln^{4+} compounds to be stable. Just before the middle of the series (Sm and Eu), and again at the end (Tm and Yb), the electron affinity of the trivalent ion is high enough to observe the 2+ state. In the solid state however, some complete series of divalent compounds are known, which display interesting trends in their electronic properties. These include the di-iodides (LnI_2), the monochalcogenides (LnS, LnSe, and LnTe), and the hexaborides (LnB_6). In the latter compounds, the structure contains B_6 octahedra linked at all their vertices, and having the formal charge $(B_6)^{2-}$, analogous to the *closo* borane anion $(B_6H_6)^{2-}$. Thus CaB_6 is non-metallic, with a gap between the valence band composed of boron–boron bonding orbitals, and the Ca 4*s* conduction band.

Many of the Ln^{2+} compounds mentioned above are metallic, and as in the case of the elements it is most unlikely that this could arise from the partially filled $4f$ orbitals, since the bandwidths of these levels are very small compared with the Hubbard repulsion energy, U. In fact, although the lowest energy configuration of most gas phase 2+ ions is $(4f)^{n+1}$, it appears that in compounds, the $(4f)^n\ (5d)^1$ configuration is more stable. There may be two reasons for this difference:

(i) The 5*d* orbitals shield the outer electrons less efficiently than do the 4*f*, and ions with the $(4f)^n\ (5d)^1$ configuration are smaller. Lattice energies of compounds with this configuration are therefore larger than ones with the $(4f)^{n+1}$ state.
(ii) The 5*d* orbitals overlap to give a considerable bandwidth. One electron near the bottom of this band therefore may have a bonding effect not possible for electrons in the 4*f* orbitals.

The metallic nature of compounds such as LaI_2 and LaB_6 is thus a result of one electron in the broad 5*d* band. Since the 4*f* configuration is that appropriate to the Ln^{3+} ion, rather than Ln^{2+}, the compounds are sometimes formulated as $Ln^{3+}\ (e^-)\ X^{2-}$, rather than as a normal divalent compound $Ln^{2-}X^{2-}$. Although such a formulation is useful in describing the electronic structure, it tends to obscure the fact that the formal oxidation state of the lanthanide is 2+.

Not all divalent compounds are metallic, however, and the changes which can occur across the series are illustrated in Fig. 5.6. The conduction band shown is derived from the Ln 5*d* orbitals, and in Fig. 5.6(a), the energy of the $(4f)^n$ to $(4f)^{n+1}$ process is above the Fermi level in this band. This is the

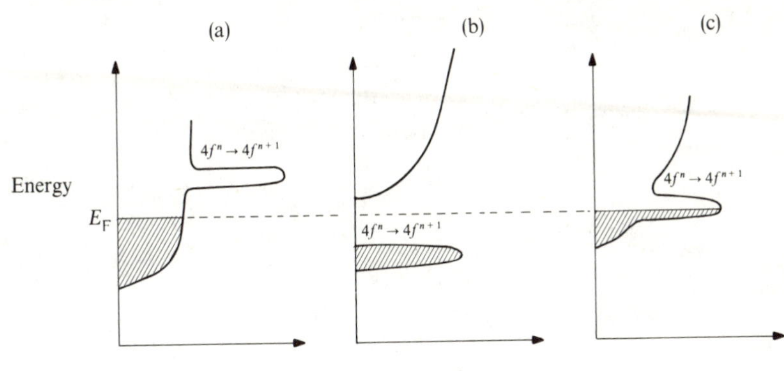

Fig. 5.6 Electronic levels in divalent compounds of lanthanides. $(4f)^n$ is again the configuration of the trivalent ion. (a) Metallic compound, with the extra electron in the 5*d* conduction band. (b) Electron captured by the 4*f* orbital: non-metallic compound. (c) Intermediate case, with mixed configuration.

situation just described, where the extra electron occupies the 5*d* band, rather than the 4*f* level. As the 4*f* binding energy increases, however, the 4*f* level may drop below the Fermi level, as shown in Fig. 5.6(b). The extra electron is now captured by the 4*f* orbitals, as in the 'divalent' metals, Eu and Yb. The compound $Ln^{2+}X^{2-}$ now has the electron configuration $(4f)^{n+1}$, and since the only electrons outside closed shells are in 4*f* orbitals, there will be no metallic conduction. It can be seen from the energies in Fig. 5.5 that this is most likely to happen in Eu and Yb, followed by Sm and Tm. These are just the elements for which non-metallic compounds such as monochalcogenides are found.

The change in the lanthanide electron configuration is also reflected in the lattice parameters, since, as noted above, the $(4f)^{n+1}$ configuration gives an ion of larger radius. The change in lattice parameters of the monosulphides (LnS), and tellurides (LnTe), are shown in Fig. 5.7. Non-metallic compounds with tellurium are found with Sm, Eu, Tm, and Yb. With sulphur, however, the extra lattice energy is sufficient to form the metallic $(4f)^n(5d)^1$ state with Tm.

The case of thulium selenide (TmSe) is particularly interesting. Its lattice parameter is intermediate between that expected for the two electron configurations, and a variety of electronic measurements show that it has a mixture of ions in the $(4f)^{12}(5d)^1$ and $(4f)^{13}$ states. In this compound, the 4*f* energy coincides with the Fermi level, as shown in Fig. 5.6(c). All the Tm ions are identical from a structural point of view, and have a kind of mixture of the two configurations. Because of the very small overlap between the 4*f* and other orbitals, however, the situation is quite different from the mixing between *s* and *d* orbitals in the band structures of the transition metals. It is better to think of the electron *hopping* rather slowly into and out of the 4*f* level. Thus

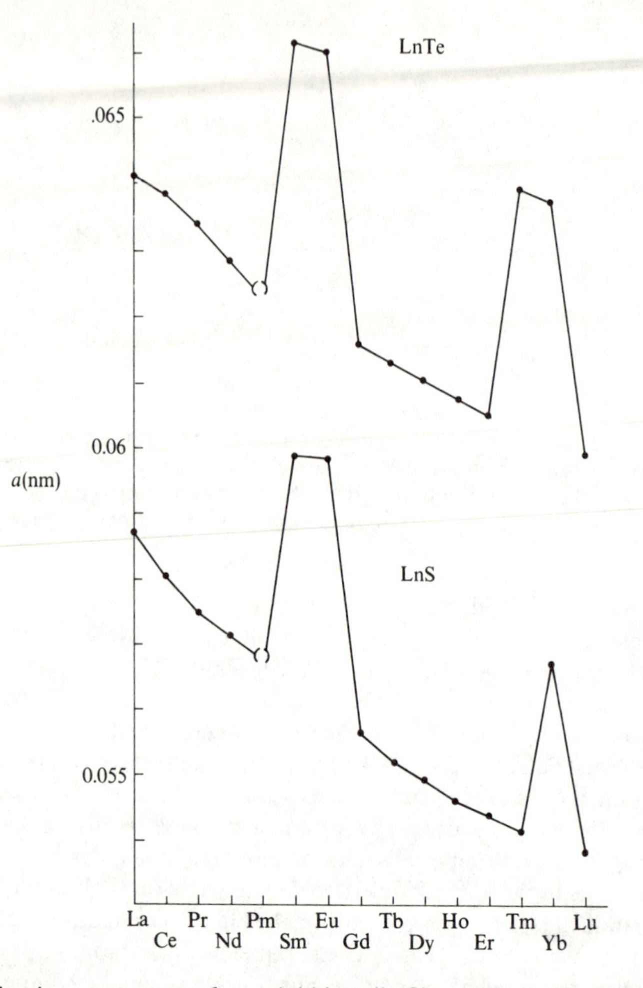

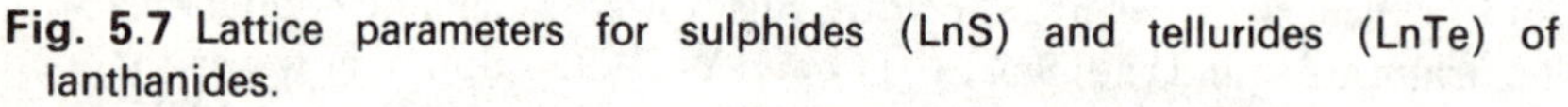

Fig. 5.7 Lattice parameters for sulphides (LnS) and tellurides (LnTe) of lanthanides.

measurements on a fast electronic time-scale (such as PES) show the two distinct configurations. TmSe has been referred to as a 'mixed valency' or 'fluctuating valency' compound, but from a chemical point of view, this is misleading, since the formal oxidation state of the lanthanide is always 2+. It would be better to call it a **mixed configuration** compound. A number of other examples are known. Samarium hexaboride (SmB_6) is one, and samarium sulphide (SmS) also shows the same behaviour under pressure, although at atmospheric pressure SmS is non-metallic, with the $(4f)^6$ configuration. Mixed configuration compounds of the lanthanides show many anomalous electronic and magnetic properties, which are not well understood.

5.3 Transition-metal compounds

The *d* orbitals in transition metals and their compounds overlap more strongly than do the 4*f* orbitals in the lanthanides. In many cases the resulting *d* bandwidth is sufficient to overcome the electron repulsion. This is the case in the metallic elements, and it is interesting to compare the photoelectron and inverse photoelectron spectra of gadolinium, shown in Fig. 5.3, with those of the transition metals shown previously in Fig. 2.6 on p. 32. With the transition metals, the occupied and empty parts of the *d* bands meet at the Fermi level, and no Mott–Hubbard splitting can be seen. This shows that U is less than W for the 3*d* levels, and band theory can be used. Although it is not strong enough to split the band, electron repulsion does have some influence, and as seen in Section 3.3.2, it gives rise to the magnetic properties of the later 3*d* elements.

In many compounds of transition metals, the *d* band is rather narrower, and quite often a Mott–Hubbard splitting is seen, with localized electrons. A rather bewildering variety of electronic properties can be observed, due to the competition between different effects. Thus changes from metallic to insulating character, and then back to metallic again, are sometimes shown by a single series of compounds. Before describing the major trends, we shall look in more detail at the first-row transition-metal monoxides, referred to in Section 3.4.4.

5.3.1 Monoxides

Monoxides (MO) with the NaCl structure are formed by the first-row transition elements (Ti, V, Mn, Fe, Co, and Ni). The first two are metallic, and the remaining ones insulating or semi-conducting. The broad features of the energy levels of a transition-metal compound have already been described, but it is interesting to apply the ideas of Chapter 3 again to this series. Figure 5.8 shows how the ionic model would be used to describe the electronic levels of a non-metallic oxide such as MnO. The levels shown are: oxygen 2*p*, which forms the valence band; metal 4*s*, which forms a broad conduction band; and metal 3*d* energies, corresponding to the two processes:

$$(3d)^n \rightarrow (3d)^{n+1} \tag{5.11}$$

and

$$(3d)^n \rightarrow (3d)^{n-1}, \tag{5.12}$$

where n is the number of 3*d* electrons in the metal 2+ ion.

The effects shown in Fig. 5.8 are the same as those discussed in Section 3.1.1. The free ion energies (a) are modified by: (b) the Madelung potential in the ionic lattice; (c) the polarization, which raises the energy of occupied levels, and lowers that of empty levels; and (d) the bandwidth coming from the overlap of

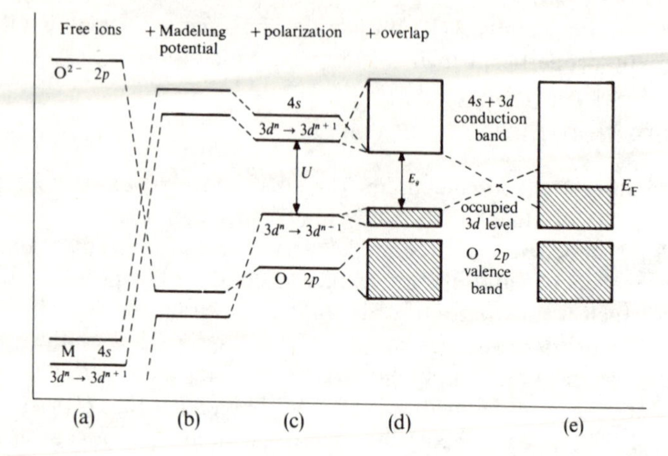

Fig. 5.8 Schematic derivation of energy levels for monoxide (MO) of an element of the first transition series. (Compare Fig. 3.1 on p. 46). (a) Free-ion levels, including the $(3d)^n$ to $(3d)^{n+1}$ and $(3d)^n$ to $(3d)^{n-1}$ energies. (b) Ions in the Madelung potential. (c) Polarization included, showing the Hubbard U for the 3d orbitals. (d) Overlap, giving narrow 3d band appropriate to a non-metallic oxide, with band gap E_g. (e) Wide d band gives a metallic compound.

orbitals. The main difference from the compounds discussed in Section 3.1 is that the 3d orbitals of the transition metal are partially filled. It is therefore necessary to consider the energies of both processes (equations 5.11 and 5.12), in which an electron is lost or added to the d shell. The Hubbard U is the difference between these two energies. In the gas phase, it would be the difference between the second and third ionization energies, around 15 eV. Polarization effects in the solid reduce the magnitude as shown in Fig. 5.8(c). Although the final value is difficult to estimate, it is probably in the range 3–5 eV for many compounds.

The situation shown in Fig. 5.8(d) is that of a non-metallic oxide, where the d bandwidth is insufficient to overcome the Mott–Hubbard splitting. Measurements by photoelectron spectroscopy show that the widths of occupied d bands in the oxides MnO to NiO are around 1 eV. The empty band may be considerably broader, since the energies of the electronic processes:

$$(3d)^n \rightarrow (3d)^{n+1}$$

and

$$(3d)^n \rightarrow (3d)^n (4s)^1$$

are quite close, and these levels probably overlap in the solid to give a single band, with both 3d and 4s character.

The band gap E_g shown in Fig. 5.8 has a different nature from that in a simple oxide such as MgO. The highest occupied level in these transition-metal oxides is composed of metal 3*d* orbitals, and the band-gap excitation is essentially a transition from an occupied 3*d* orbital on one cation, into an empty orbital on a neighbouring one. In MgO, the band gap is between a valence band made up of oxygen 2*p* orbitals, and the magnesium 3*s* conduction band. Transitions from the oxygen levels can also be identified in the transition-metal oxides, and occur at energies well above the main absorption edge. Measurements by electronic absorption spectroscopy and photoelectron spectroscopy give experimental estimates of the energies of the different levels in Fig. 5.8(d), and values for some of the oxides are shown in Fig. 5.9. Although this series is less complete than that of the 4*f* levels in Fig. 5.5, it shows the same trends. There is a stabilization of the 3*d* levels with increasing nuclear charge (Fe, Co, Ni) and a break after the half-filled shell (Mn, $(3d)^5$). The high position (which implies low binding energy) of the occupied 3*d* level in FeO is of considerable chemical significance. All these oxides can be made with a stoichiometric deficiency in the metal. The structures show cation vacancies, and the missing positive charge is compensated by oxidation of some M^{2+} to M^{3+}. This is much easier to do in FeO than in the other non-metallic oxides, and in fact that 'FeO' wustite phase is never stoichiometric, but is always deficient in iron. The relative ease of oxidation of Fe^{2+} is of course apparent in its aqueous chemistry, and has essentially the same cause.

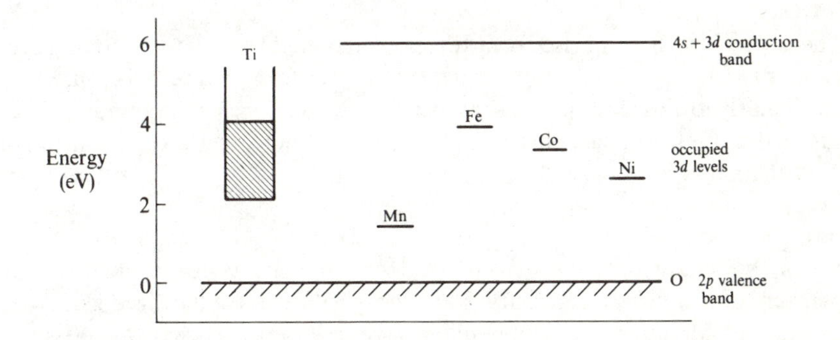

Fig. 5.9 Electronic energy levels of some 3*d* monoxides, deduced from spectroscopic measurements. The energy zero has been taken as the top of the oxygen 2*p* valence band.

Figure 5.8(e) shows what happens when the *d* band becomes broader. The occupied and empty parts now overlap, giving a metallic compound, with a Fermi level in the middle of a partially filled band. This is the same effect as in the simpler case illustrated in Fig. 5.2, where the Hubbard sub-bands overlap in energy when the bandwidth is large. The resulting energy level scheme is applicable to the metallic oxides TiO and VO. The experimental energy levels of TiO deduced from photoelectron spectra are also shown in Fig. 5.9. The extra $3d$ bandwidth in these compounds, as mentioned in Section 3.4.3, arises partly from the more diffuse d orbitals earlier in the series, but is also related to the high concentration of vacancies. It has been suggested that another factor giving a larger $3d$ bandwidth is the proximity of the wide $4s$ band. Detailed calculations do not support this idea however, and the lower part of the conduction band is mostly $3d$ in composition. The metallic $4d$ compound NbO is rather similar to TiO and VO, except that the vacancies are ordered, rather than randomly arranged.

5.3.2 General trends

It would be impossible here to give a complete survey of transition-metal compounds. It is useful however to look at some of the general trends observed in their electronic structure. As with the monoxides, anything which increases the d bandwidth, or decreases the repulsion energy U, is more likely to lead to delocalized electrons. Some of the important factors are as follows:

(i) Position in the series

Larger d orbitals give a broader band and smaller U. This happens with elements earlier in a transition series (as in the monoxide series), or with elements of the second or third transition series.

(ii) Oxidation state

The d bandwidth can arise from direct metal–metal overlap, or by indirect covalent bonding via an intervening anion. The former case is favoured by particularly low oxidation states (as in TiO and metal-rich compounds such as ZrCl), the latter by a particularly high one (as in ReO_3, or $LaNiO_3$, which are both metallic).

(iii) The anion partner

When the bandwidth arises from covalency, it is likely to be increased by partners of low electronegativity. Thus most halides have localized electrons; oxides and sulphides are intermediate, with a variety of behaviour possible; heavier chalcogenides, phosphides, etc., are more often metallic.

(iv) Electron configuration

We saw with the lanthanides that the exchange stabilization of parallel spin electrons gives an extra contribution to U at the half-filled shell. The same effect is seen in the transition series: $(3d)^5$ compounds of Mn^{2+} are noticeably less likely to be metallic than those of nearby elements. For example, many sulphides (MS) have the NiAs structure, where there is metal–metal bonding and often metallic character. (Many of these form phases of variable stoichiometry, like TiO and VO.) However, MnS has the NaCl structure, and is non-metallic.

(*v*) *Other cations*
Ternary compounds, such as the perovskite oxides (ABO_3) also have another cation present. When this is an A metal, it will have empty orbitals considerably higher in energy than the transition metal *d* levels, and its influence on the electronic structure will generally be small. Post-transition B metals, however, have *s* orbitals, which may be filled, in the same energy range as the *d* band. Interaction between the two cations, either by direct overlap or indirectly via an anion, can broaden the *d* band. Thus in $Y_2Ru_2O_7$ the Ru^{4+} has a localized $(4d)^4$ configuration, but $Bi_2Ru_2O_7$ is metallic, and PES measurements suggest that this is due to some involvement of the 6*s* electrons present in Bi^{3+}.

There is another factor that can help to give rise to metallic behaviour. The simple discussion of the Hubbard model in Section 5.1 assumed that we had a half-filled band, although any integral number of electrons per atom would lead to the same conclusion. But if some extra electrons are added, they can move from atom to atom without any additional electron repulsion (see Fig. 5.10). The same is true if electrons are removed from the lower sub-band. It would seem therefore that a non-stoichiometric compound, or one with mixed valency, should be metallic, irrespective of the relative magnitudes of U and W. One example is Fe_3O_4, which does indeed show high conductivity at normal temperatures. This can be understood in terms of easy electron transfer from Fe^{2+} to Fe^{3+}: both ions are in equivalent octahedral sites in the inverse spinel lattice. However, many mixed valency and non-stoichiometric compounds are *not* metallic, and Fe_3O_4 itself becomes a semiconductor below 135 K. The reason is that compounds with narrow bands are also subject to lattice distortions which tend to trap the extra electrons or holes. These effects are explained in more detail in Chapter 6.

Finally, it must be remembered that transition-metal compounds can be non-metallic for reasons that have nothing to do with electron localization. Section 3.4.2 explained how the *d* band can be split by ligand field effects, and

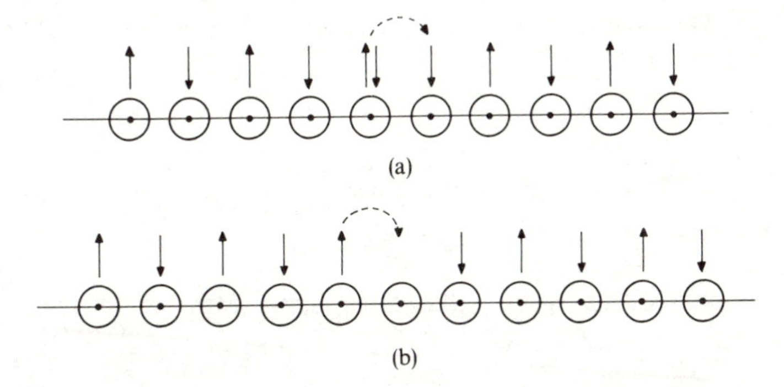

Fig. 5.10 Motion of extra electrons (a) or holes (b) is possible without incurring additional electron repulsion.

how certain electron configurations can give rise to filled bands. The band gaps in these compounds may sometimes have a contribution from electron repulsion, but they are primarily due to chemical bonding effects, similar to those treated by the conventional band model.

5.3.3 Properties of localized electrons: ligand field effects

The properties of compounds with localized *d* electrons are also influenced strongly by ligand field splittings. These effects are much more important than in the lanthanide series, because the partially filled *d* orbitals interact more with neighbouring atoms than do the 4*f* orbitals. As explained in Section 3.4.2, the splitting of *d* orbitals is a consequence of their different bonding interactions with the surrounding ligands. Orbitals that point directly towards the ligands form σ bonds. The ligand electrons are accommodated in the bonding molecular orbitals formed, and so the 'metal *d*' electrons have to occupy the higher-energy antibonding set. For the common case of octahedral coordination shown previously in Fig. 3.17 (p. 74), there are two σ antibonding *d* orbitals, d_{z^2} and $d_{x^2-y^2}$, normally referred to by their symmetry label e_g. The other three *d* orbitals (d_{xz}, d_{yz}, and d_{xy}), known as the t_{2g} set, do not point directly towards the octahedral ligands, and so only form π antibonding combinations. Since the σ interaction is stronger, the e_g set is higher in energy.

As the *d* shell is filled across the transition series, the orbitals will be occupied in an order that depends on the relative magnitude of the ligand field splitting Δ, and the intra-atomic exchange interaction which favours parallel spins. Section 3.4.2 discussed cases where the electrons were paired up in the lower set of *d* orbitals. These *low-spin* configurations occur if Δ is sufficiently large. Low-spin configurations are sometimes found for localized electrons, as in the $(t_{2g})^4$ state of Ru^{4+} in $Y_2Ru_2O_2$. For the majority of localized electron configurations in halides and oxides, however, the exchange term is more important, and electrons will be arranged in the *d* orbitals so as to have as many as possible with parallel spin. For example, in MnO the electron configuration is $(t_{2g})^3\,(e_g)^2$. The high-spin electron configurations for an octahedral field are shown in Fig. 5.11.

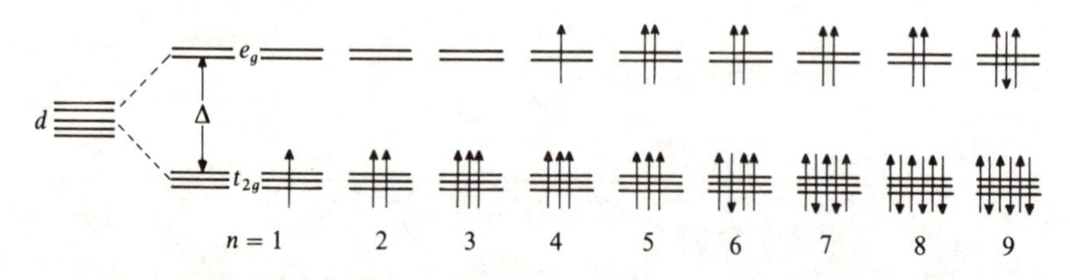

Fig. 5.11 High-spin electron configurations for transition-metal ions in octahedral coordination.

One very well-known consequence of the ligand field splitting is the occurrence of electronic transitions between the different *d* orbitals. Ligand field spectra can be observed in solids with localized *d* electrons, and generally appear quite similar to those in isolated transition-metal complexes. Figure 5.12, for example, shows the optical absorption spectrum of nickel oxide (NiO), compared with that of the complex ion $Ni(H_2O)_6^{2+}$ measured in aqueous solution. The three bands in the aqueous spectrum are the ligand field transitions, and these appear to be very similar in the solid. This is to be expected, since in both cases, nickel is surrounded by six oxygen atoms.

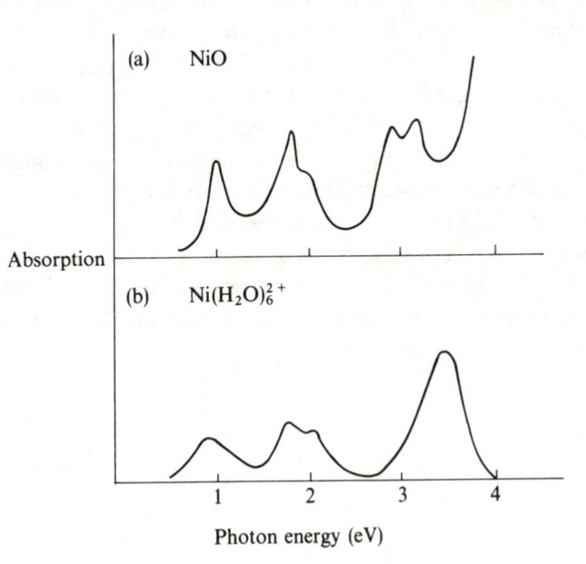

Fig. 5.12 Optical absorption spectra of (a) NiO and (b) $Ni(H_2O)_6^{2+}$, showing the three spin-allowed ligand field bands.

A detailed analysis shows that the ligand field bands in octahedral Ni^{2+} arise from the different arrangements possible for the eight electrons in the t_{2g} and e_g orbitals. The ground-state electron configuration is $(t_{2g})^6\,(e_g)^2$, and group-theoretical arguments show that the two electrons in the e_g orbitals can be arranged in different ways, to give the spectroscopic states:

$$^3A_{2g} \quad ^1A_{1g} \quad \text{and} \quad ^1E_g$$

The first of these has the two electrons with parallel spins, and is the ground state. Various excited states come from the excited configurations:

$$(t_{2g})^5\,(e_g)^3\text{: } ^3T_{1g},\ ^3T_{2g},\ ^1T_{1g},\ \text{and}\ ^1T_{2g}$$

$$(t_{2g})^4\,(e_g)^4\text{: } ^3T_{1g},\ ^1A_{1g},\ ^1E_g,\ \text{and}\ ^1T_{2g}$$

The normal spin selection rule shows that the triplet excited states will give by far the most intense transitions in the spectrum, and these form the three bands observed. The energies of the states depend not only on the ligand field splitting Δ, but also on the different electron repulsion energies resulting from the different arrangements of electrons in the *d* shell.

The similarity between the solid state and the complex-ion spectra emphasizes the highly localized nature of the 3*d* electrons in NiO. Other measurements suggest a *d* bandwidth of about 1 eV. This bandwidth is not seen in the absorption spectra, where the width of the peaks, in the solid as in solution, is almost entirely due to vibrations excited with the electronic transition. The ligand field excited states are localized, in fact, by exactly the same electron repulsion effects that operate in the ground state. Ligand field excitations can be regarded as 'Frenkel excitons' similar to those found in molecular solids, and discussed in Chapter 7.

At higher energies, the spectra of NiO and the hexaaquo complex are quite different. The strong absorption edge at 3.8 eV in the solid represents the band gap shown previously in Figs 5.8 and 5.9.

Ligand field splittings can also have structural consequences, because the bonding influence of the *d* electrons changes as the orbitals are filled. Figure 5.13 shows the metal–ligand distances in some M^{2+} compounds of the

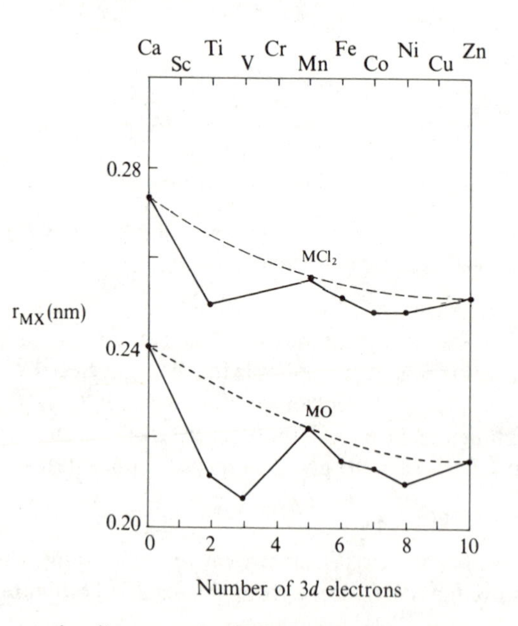

Fig. 5.13 Inter-atomic distances in MO and MCl_2 compounds of the first transition series. (From C. S. G. Phillips and R. J. P. Williams, *Inorganic Chemistry,* Oxford University Press, 1965.)

first transition series. The overall trend is a contraction of the metal ions as the nuclear charge increases. Superimposed on this however is a pattern that reflects the orbital occupancy. There is an increase of ionic radius between d^3 and d^5, and between d^8 and d^{10}. All the M^{2+} ions have high-spin electron configurations in octahedral sites. As seen by a comparison with Fig. 5.11, the increase of M–X distance occurs for the cases where electrons are added to the more antibonding e_g orbitals.

Such a plot of average distance obscures another very interesting consequence of the ligand field splitting. Whenever a set of degenerate orbitals is unevenly occupied by electrons, the charge distribution must be unsymmetrical, and produces a force that tends to distort the geometry of the surrounding ligands. This is the **Jahn–Teller effect**. For octahedral geometries, an uneven orbital occupancy happens when the t_{2g} orbitals have one, two, four, or five electrons, or with one or three electrons in e_g. In practice, distortions produced by electrons in the t_{2g} orbitals are small and often difficult to observe, but the effect is generally pronounced with e_g orbitals. The appropriate high-spin electron configurations are:

$$d^4\,(t_{2g})^3\,(e_g)^1$$

and

$$d^9\,(t_{2g})^6\,(e_g)^3.$$

The Jahn–Teller distortion is generally seen as a tetragonal elongation of the octahedron, so that there are four short bonds and two long ones. Figure 5.14 illustrates this distortion, and shows how it alters the ligand field energies of the d orbitals. The d_{z^2} orbital is lowered at the expense of $d_{x^2-y^2}$, and the electrons are arranged so as to give a net stabilization.

The commonest examples of d^4 ions are Cr^{2+} and Mn^{3+}, and in non-metallic compounds these ions invariably produce a lattice distortion. The same is true with Cu^{2+}, d^9. In the solid state the distortions around each ion must be oriented in such a way as to fit together in the crystal lattice, and the result is known as a **cooperative Jahn–Teller distortion.**

A good example of a cooperative Jahn–Teller distortion can be seen in rubidium chromous chloride (Rb_2CrCl_4). The structure is based on that of K_2NiF_4, and has layers separated by the alkali ions. The transition metal has an octahedral coordination of halides, with four in the plane (shown in Fig. 5.15), and one above and below (not shown). In Rb_2CrCl_4 the planes are distorted, so that there are two long Cr–Cl distances, and two short. Each out-of-plane Cl has a short separation, so that the overall coordination round Cr is tetragonal, as in Fig. 5.14. A very similar distortion is seen in K_2CuF_4. Cooperative Jahn–Teller distortions of a related kind also occur in oxides, such as the perovskite lanthanum manganate ($LaMnO_3$). As will be seen in the next

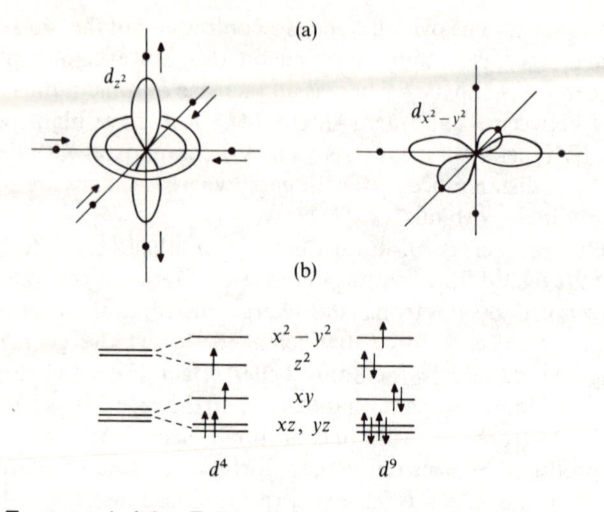

Fig. 5.14 Tetragonal Jahn–Teller distortion of a six-coordinate complex. (a) Geometries of the d_{z^2} and $d_{x^2-y^2}$ orbitals in the distorted environment. (b) Energy splitting produced by the distortion, showing $(d)^4$ and $(d)^9$ electron configurations.

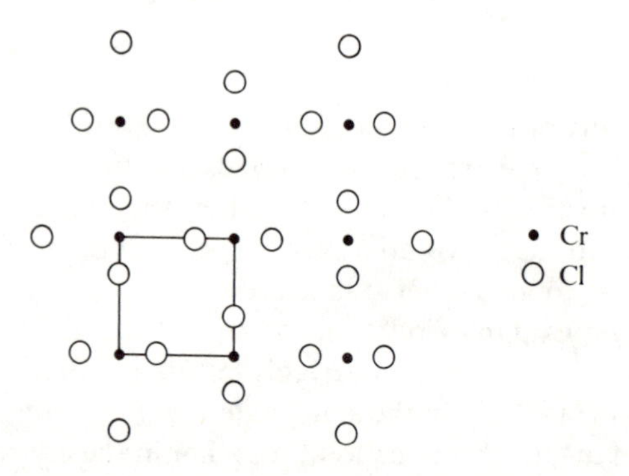

Fig. 5.15 Cooperative Jahn–Teller distortion in Rb_2CrCl_4. Only chlorides in the plane are shown; each Cr^{2+} has Cl^- above and below it. The distortion is exaggerated for clarity. (From P. Day, M. T. Hutchings, E. Janke, and P. J. Walker, *J. C. S. Chem. Comm.*, (1979), 711.)

section, these distortions can have an important influence on the magnetic properties of these compounds.

5.3.4 Magnetic properties of localized electrons

Unpaired electrons in localized states have magnetic properties very different from those of metals. In Section 3.3.1, we saw how the **Pauli susceptibility** of a simple metal is independent of temperature. Isolated complexes with unpaired electrons, however, have a magnetic susceptibility obeying the **Curie Law**:

$$\chi = C/T. \tag{5.13}$$

The Curie constant C depends on the magnetic moment μ of the electrons:

$$C = N\mu_0\mu^2/(3k). \tag{5.14}$$

In the lanthanides, the magnetic moment always has an orbital contribution resulting from the angular momentum of the $4f$ electrons. In the transition series however, the orbital motion of d electrons is largely quenched by ligand field effects, and it is often a good approximation to use the *spin-only* formula:

$$\mu = g\{S(S+1)\}^{1/2}\,\mu_B. \tag{5.15}$$

In this equation μ_B is the Bohr magneton, and S the spin quantum number of the complex. The g factor is close to 2 for free electrons, although some deviations occur from spin–orbit coupling.

The Curie Law is only valid for magnetic ions that are essentially isolated from one another. In a solid, the ions always interact to some extent, and the magnetic properties are more complex. Figure 5.16 shows three ways in which the susceptibility can vary with temperature. The *paramagnetic* case corresponds to no interaction between ions, so that the Curie Law is obeyed. In a *ferromagnetic* solid, the susceptibility rises faster as the temperature is lowered, and at the *Curie temperature*, T_C, the magnetic interactions between the ions

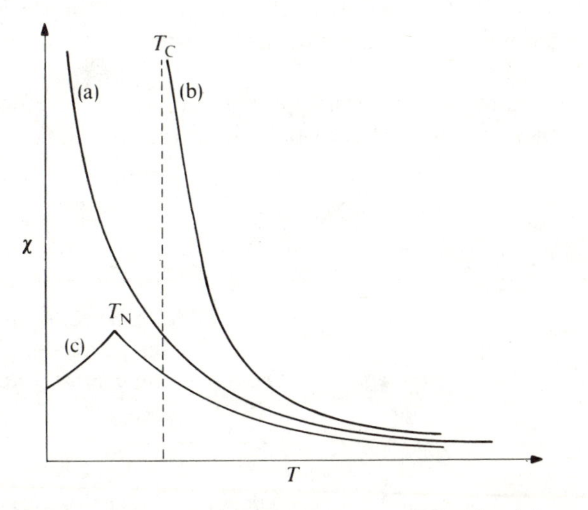

Fig. 5.16 Schematic variation of magnetic susceptibility for (a) paramagnetic, (b) ferromagnetic, and (c) antiferromagnetic solids.

cause the spins to align parallel to each other. In the *antiferromagnetic* case, the alignment occurring at the *Néel temperature*, T_N, leads to a net cancellation of the moment on different ions, so that the susceptibility is lowered. In the simplest form of antiferromagnetic alignment, alternate ions in the lattice have their magnetic moment pointing in opposite directions, but more complicated arrangements are possible.

Well above the ordering temperatures, the magnetic susceptibility is given by the **Curie–Weiss** formula:

$$\chi = C/(T-\theta). \tag{5.16}$$

The Weiss constant θ should be positive for a ferromagnet, and negative in the case of antiferromagnetic alignment.

When solids contain more than one type of magnetic ions, more complex patterns of magnetic ordering can be found. For example **ferrimagnetism** is a result of an anti-parallel alignment of spins on ions of different type, so that there is no net cancellation of magnetic moment as with antiferromagnetism.

Tables 5.1 and 5.2 show the magnetic properties of a selection of transition-metal compounds, all with localized 3*d* electrons. The magnetic moments are generally in reasonable agreement with the spin-only formula of equation 5.15. Some of the deviations are due to small orbital contributions; in other cases the magnetic moments in the tables may not be very reliable, as the Curie–Weiss law is not obeyed well in solids with high ordering temperatures. Although magnetic susceptibility measurements give an indication of the type of magnetic ordering, they do not show the detailed arrangement of spins. The best technique for this is neutron diffraction, which utilizes that fact that neutrons are scattered by unpaired electrons. Figure 5.17 shows the alignment of magnetic moments below the Néel temperature in the antiferromagnetic solid MnO.

Magnetic ordering lowers the entropy of a solid, and is therefore opposed by the thermal agitation that tends to randomize the spin directions. Thus the

Table 5.1

Magnetic properties of some solids: Ferromagnetic compounds

Compound	*Magnetic ion*	*Spin* (S)	*Observed magnetic moment* (μ/μ_B)	*Curie temperature,* T_C (K)
Rb_2CrCl_4	Cr^{2+}	2	5.8	57
K_2CuF_4	Cu^{2+}	1/2	1.8	6
$La_{0.7}Sr_{0.3}MnO_3$	Mn^{3+}, Mn^{4+}	2, 3/2	3.7	350

Table 5.2

Magnetic properties of some solids: Antiferromagnetic compounds

Compound	*Magnetic ion*	*Spin* (S)	*Observed magnetic moment* (μ/μ_B)	*Weiss constant* θ (K)	*Néel temperature,* T_N (K)
MnO	Mn^{2+}	5/2	5.5	−417	122
MnF_2	Mn^{2+}	5/2	6.0		67
FeO	Fe^{2+}	2	7.1*	−507	198
CoO	Co^{2+}	3/2	5.0*	−300	292
NiO	Ni^{2+}	1	4.6*	−2000	530
NiF_2	Ni^{2+}	1	3.6	−116	83

* Value probably unreliable, as Curie–Weiss law is not well obeyed.

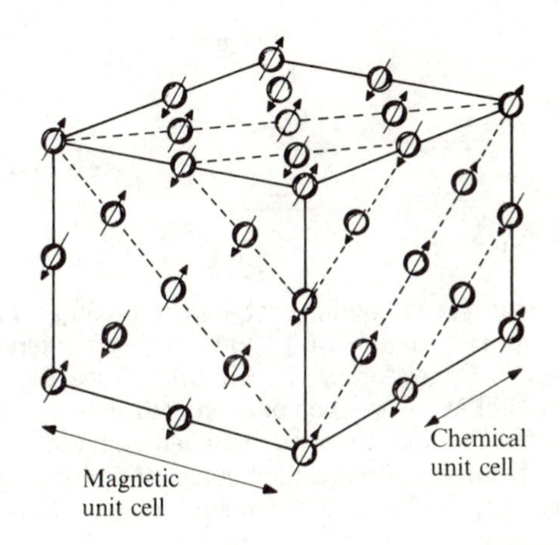

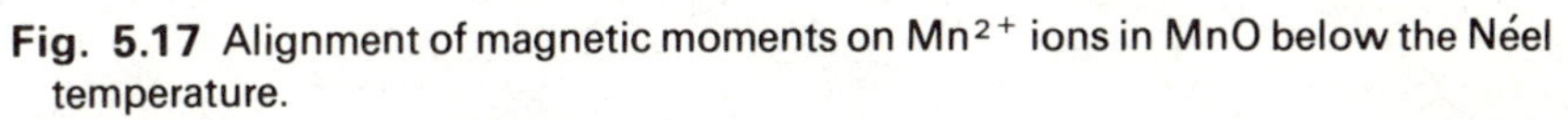

Fig. 5.17 Alignment of magnetic moments on Mn^{2+} ions in MnO below the Néel temperature.

ordering temperature depends on the strength of the interaction between neighbouring ions. The Néel temperature of NiO is 530 K, but this is exceptionally high, and many magnetic solids have ordering temperatures below 10 K. Although there is a direct through-space dipole–dipole interaction between magnetic ions, calculations show that in most cases this effect is much too weak to account for the observed ordering. More important is the **exchange interaction**, which arises from chemical bonding effects similar to those giving the width of the *d* band. The exchange interaction can come from direct overlap between the d orbitals holding the un-paired electrons. As seen previously however, interaction between transition-metal ions is often indirect, and involves covalent bonding with intervening ligand atoms. Magnetic interaction of this kind is known as **superexchange**. Some of the effects possible are illustrated in Fig. 5.18.

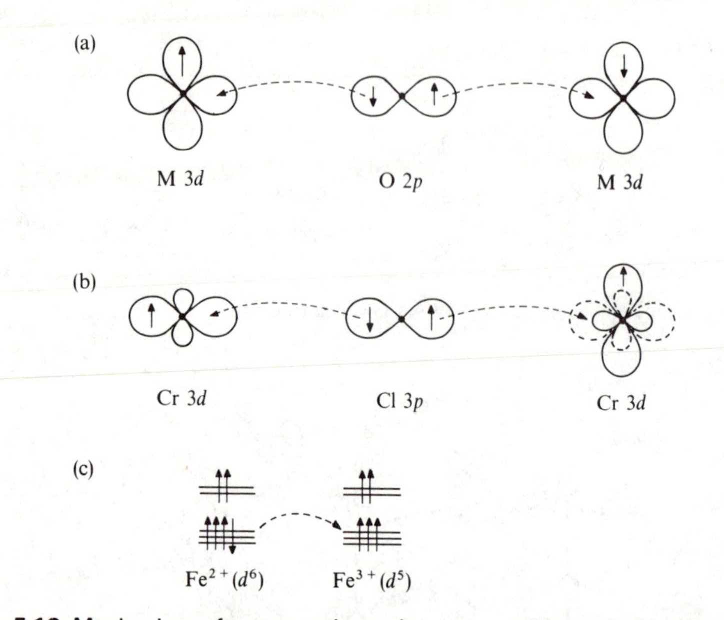

Fig. 5.18 Mechanisms for magnetic exchange coupling. (a) Superexchange giving antiferromagnetic alignment of ions with an intervening anion. (b) Cooperative Jahn–Teller distortion in Rb_2CrCl_4 (see Fig. 5.15) results in coupling of a filled orbital (———) on one ion with an empty orbital (- - - -) on a neighbouring one. The net coupling between spins in occupied orbitals is ferromagnetic. (c) Double exchange: the transfer of a minority spin electron from Fe^{2+} to Fe^{3+} is only possible if the majority spins have a ferromagnetic alignment.

The top diagram shows the situation in the monoxides MO, where two metal atoms are separated by oxygen. Overlap between the metal d orbitals and the oxygen $2p$ leads to some covalent mixing, and has the effect of partially transferring an electron from the oxygen to the metal orbitals. If metal A has an unpaired electron with spin up, the exclusion principle shows that only a spin-down electron can be transferred to the same orbital. Bonding to the metal atom B, on the other side, must therefore involve the spin-up electron from oxygen. This in turn is only possible if metal B has its unpaired electron in the spin-down state. The shared covalency of the two metal atoms therefore leads to an antiferromagnetic alignment of their moments. The superexchange mechanism predicts that the strongest interaction will be an antiferromagnetic one between the next-nearest neighbour metal ions, separated by an oxide ion. The magnetic structure of MnO shown in Fig. 5.17 supports this conclusion, as all next-nearest-neighbour pairs of magnetic ions have oppositely aligned spins. The influence of covalency can be seen from the order of Néel temperatures in Table 5.1, increasing in the sequence MnO, FeO, CoO, and NiO. The increasing effective nuclear charge along the transition series leads to a greater metal–oxygen mixing, and hence to a stronger superexchange interaction.

Although superexchange more often gives an antiferromagnetic coupling of spins, there are situations where it favours ferromagnetic alignment. An example is seen in rubidium chromous chloride (Rb_2CrCl_4) which was discussed in the previous section as an illustration of the cooperative Jahn–Teller effect (see Fig. 5.15). As shown in Fig. 5.18(b), the occupied 3*d* orbitals on adjacent Cr^{2+} ions point in different directions as a result of the distortion. The filled halogen orbital which overlaps with a half-filled *d* orbital on one side, has an *empty d* orbital on the other. The same argument as before shows that if the first metal ion has a spin-up electron, there will be a net transfer of spin-up electron density into the empty orbital of the other. The intra-atomic exchange interaction between electrons in different orbitals on this second atom then leads to a lower energy if the occupied orbital also has a spin-up electron. There is therefore a ferromagnetic interaction between the neighbouring Cr^{2+} ions.

It has been mentioned that electron transfer between metal ions is facilitated in mixed-valency compounds, such as Fe_3O_4. This transfer can lead to another type of magnetic exchange interaction, known as **double exchange**. The easiest electron to move from a high-spin Fe^{2+} to a neighbouring Fe^{3+} is the minority-spin electron, since it leaves the remaining electrons with parallel spin (see Fig. 5.18(c)). Such a transfer is only possible, however, if the Fe^{3+} which is to receive the electron has its own spins aligned parallel. The delocalization of electrons between the Fe ions leads to a lowering of energy, and will favour a ferromagnetic ordering of the magnetic moments. This is observed for the Fe^{2+} and Fe^{3+} ions in octahedral sites in Fe_3O_4. There is another set of Fe^{3+} ions in tetrahedral sites, but these have orbitals at a different energy, and do not participate in the electron transfer process. Superexchange leads to an antiferromagnetic alignment of spins on the tetrahedral iron with respect to those in octahedral positions. Fe_3O_4 is thus ferrimagnetic.

Interesting magnetic properties can also be found in some transition-metal compounds where the *d* band is wide enough to give delocalized electrons. For example, the metallic oxide CrO_2 is ferromagnetic, with a Curie temperature $T_C = 392$ K. As shown for the transition elements in Section 3.3.2, magnetic ordering can arise in a metal if the band is sufficiently narrow. Then it becomes favourable to align the electron spins on one atom, at the expense of losing some bonding energy. The magnetic properties of many metallic compounds cannot be understood by a simple band picture. For example, the magnetic susceptibility often varies with temperature, following the Curie–Weiss law more closely than the Pauli one. Even when magnetic ordering does not occur, this kind of behaviour is generally a sign that the *d* bands are rather narrow, and that electron repulsion effects are important.

5.4 Other theories of interelectron repulsion

The Hubbard model provides a useful way of thinking about electron repulsion in solids with narrow bands of *f* and *d* electrons. It concentrates, however, entirely on short-range effects within one atom. There are situations where longer-range repulsion is important. For example, the Hubbard theory cannot describe very well the actual transition from localized to itinerant electrons. The two sections which follow look briefly at very different approaches to the problem of electron localization.

5.4.1 Wigner crystallization

The importance of long-range electron repulsion effects in solids was first emphasized by Wigner. He showed that at very low electron densities, the free-electron gas of a simple metal is unstable, and that the electrons will 'crystallize' to form a state where each electron is localized on a body-centred cubic lattice. It is this structure that minimizes the repulsion between the electrons. The electron density required for Wigner crystallization to occur is difficult to estimate precisely, but it is certainly well below that found in most metallic elements. It has been suggested that long-range repulsion of this kind might play a role in mixed-valency compounds such as Fe_3O_4. Below 120 K this compound is not metallic, and contains localized Fe^{2+} and Fe^{3+} ions on distinguishable sites. The localization gives a regular spacing of the extra electrons, and must certainly be helped by long-range repulsions. It is likely, however, that the lattice distortion effects described in Chapter 6 are also important in mixed-valency compounds such as Fe_3O_4.

Although there is no well-authenticated example of a Wigner crystallization in three dimensions, recent experiments on electrons confined to two-dimensional regions in junctions between doped semiconductors do seem to show localization of the Wigner type.

5.4.2 The polarization catastrophe model

Another approach to the long-range electrostatic interaction of electrons is based on the dielectric properties of solids, which were discussed in Section 2.4 (p. 39). Equation 2.6 gave the dielectric function for a model solid, composed of N oscillators per unit volume, each having a fundamental frequency ω_0:

$$\varepsilon(\omega) = 1 + \frac{(Ne^2/\varepsilon_0 m)}{\omega_0^2 - (Ne^2/3\varepsilon_0 m) - \omega^2 + i\omega/\tau}. \tag{5.17}$$

The term $(Ne^2/3\varepsilon_0 m)$ in the denominator comes from the mutual polarization of the groups in the solid, which act to reinforce the effect of any applied electric field, and to reduce the excitation frequency. At zero frequency ω, this equation becomes:

$$\varepsilon_s = 1 + \frac{(Ne^2/\varepsilon_0 m)}{\omega_0^2 - (Ne^2/3\varepsilon_0 m)} \tag{5.18}$$

Where ε_s is the static dielectric constant. When the oscillators are packed together at a certain critical concentration N_c, given by:

$$1/N_c = e^2/(3\varepsilon_0 m\omega_0^2) \tag{5.19}$$

the denominator in equation 5.18 becomes zero, showing that ε_s is infinite. This is called a **polarization catastrophe**. At the critical concentration, the excitation frequency in the solid is reduced to zero. If this happens for electronic excitations, it must mean that the solid is metallic. A polarization catastrophe can also occur in the part of the dielectric function due to atomic displacements (see Section 3.2.3 on p. 61). The atoms in the solid are then displaced from their normal lattice positions, in such a way as to give a net dipole moment. Solids in which this happens, such as barium titanate ($BaTiO_3$) are called **ferroelectric**, and have important applications, for example in capacitors for electronic circuits. Our interest here, however, is only in the electronic polarization catastrophe, which predicts a transition from non-metallic to metallic properties.

The model of a single oscillator gives a rather crude picture of the electronic behaviour of an atom or molecule, but a connection with real systems may be made by noting that the right hand side of equation 5.19 is the electronic polarizability of our model oscillator. The equation can be written in a different form, using the **molar polarizability** or **molar refractivity**, R, and the molar volume in the solid, V. The criterion for a polarization catastrophe is then:

$$R/V = 1. \tag{5.20}$$

As its name suggests, the molar refractivity may be estimated from refractive index measurements of atoms or molecules under dilute conditions, for example in the gas phase or solution. Equation 5.20 predicts that for any atom or molecule, there is a certain critical molar volume, below which a metallic solid should be formed. Edwards and Sienko have applied this idea to the elements in their solid state, and have produced the plot shown in Fig. 5.19. It is remarkable that, apart from one or two borderline cases, the criterion $R/V = 1$

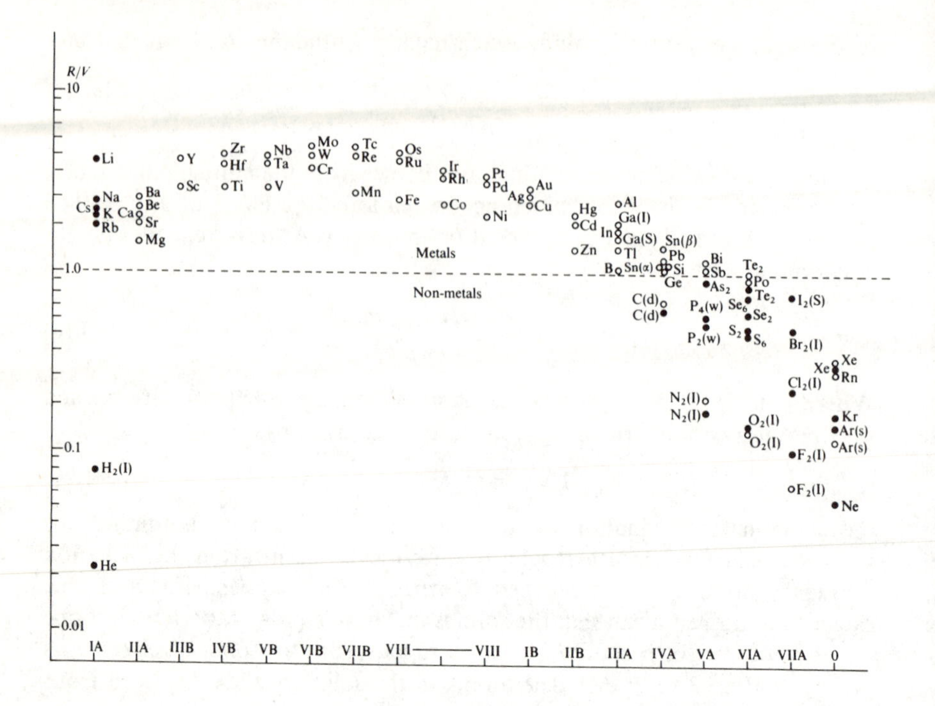

Fig. 5.19 The Periodic Table of Elements. Elements with $R/V > 1$ should be metallic, those with $R/V < 1$ non-metallic. (From P. P. Edwards and M. J. Sienko, *Chem. Brit.* (1983), 39.)

entirely separates metallic elements from the non-metallic ones. The trend across the Periodic Table is easily explained in this theory as a consequence of the increasing effective nuclear charge experienced by valence electrons. They become progressively more tightly bound, so that the atomic polarizability decreases. The opposite trend is apparent down the non-metal groups; here the increase of principal quantum number leads to an increase of polarizability, so that R/V rises towards the metallic borderline.

The polarization catastrophe model concentrates entirely on the electrostatic interaction between atoms in the solid. It is the mutual polarization of valence electrons on different atoms that forces them to become free at a certain critical density. This picture is very different from the one used so far, where we have considered the formation of bands by the overlap of orbitals. In fact the overlap of atoms is not even included in the polarization catastrophe calculation. It is likely that there is some deeper connection between the two approaches, but this has not yet been made.

5.4.3 Metal–insulator transitions at high and low densities

The most important prediction of the polarization catastrophe theory is that any solid should become metallic at a sufficiently high density, where the molar volume is small enough to satisfy equation 5.20. Although in many cases the pressures required to give such densities are inaccessible in the laboratory, this prediction is almost certainly correct. Solid hydrogen, for example, is believed to be metallic at pressures above about 2 Mbar (2×10^{11} Pa). The planet Jupiter probably contains a central core of metallic hydrogen. Transitions to the metallic state have been studied more extensively for solids that are closer to the borderline. A well-known example is iodine, which at normal pressures forms a molecular lattice. Figure 5.20(a) shows how the electrical conductivity increases with pressure, and reaches a metallic value at 170 kbar. At the same time, the band gap measured spectroscopically decreases gradually to zero. In some other cases, more abrupt changes are observed. In Fig. 5.20(b) it is shown how the conductivity of elemental silicon and germanium change with pressure. The first-order transition to a metallic state in these elements is accompanied by a change in crystal structure, from the normal diamond form, to the metallic structure of white tin.

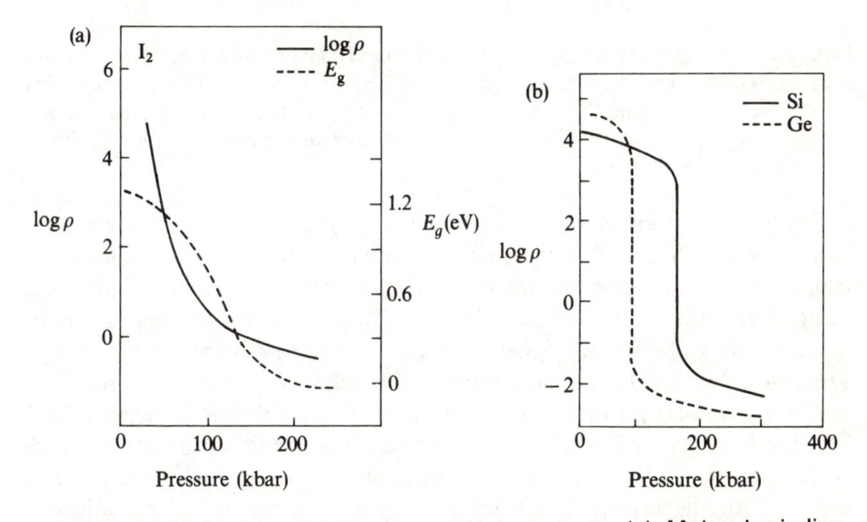

Fig. 5.20 Transitions to metallic state at high pressure. (a) Molecular iodine, showing electrical resistivity and band gap. (b) Electrical resistivity of silicon and germanium. (From H. G. Drickamer and C. W. Frank, *Electronic transitions and the high-pressure chemistry and physics of solids* Chapman and Hall, 1973.)

The converse prediction is that any solid that is metallic under normal conditions should cease to be so as the density is lowered. Such a reduction in density can be achieved in various ways. For example, it is possible to dilute metal atoms in an inert matrix, such as solid argon at low temperatures. It is found that a certain critical concentration is required for metallic conductivity to occur. Experiments have also been performed on metal vapours at temperatures above the critical point, where the density can be varied continuously. Figure 5.21 (a) shows the conductivity of caesium, measured as a function of density in this way. The normal solid density of the metal is 1.9 $g\,cm^{-3}$, and it can be seen that the conductivity drops rapidly at about half this density. This is in good agreement with the predictions of equation 5.20.

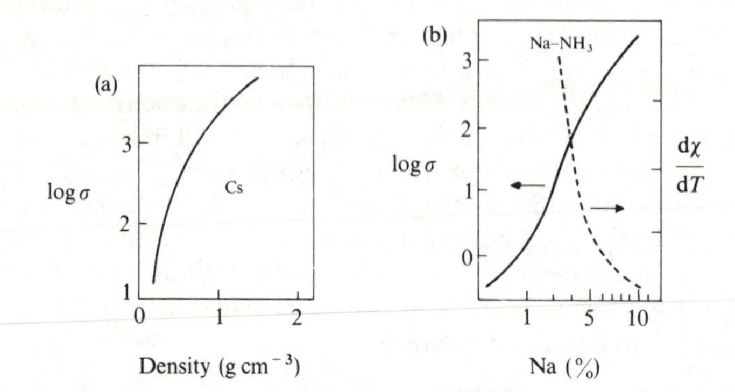

Fig. 5.21 'Expanded metals'. (a) Conductivity of caesium as a function of density, measured at temperatures above the critical point. (b) Conductivity and temperature coefficient of magnetic susceptibility for sodium in liquid ammonia. (From J. C. Thompson, *Electrons in liquid ammonia,* Oxford University Press, 1970.)

The expanded metals best known in chemistry are those formed by dissolving alkali metals in liquid ammonia. At low concentrations, the properties of the solutions are generally interpreted in terms of electrons trapped by solvating ammonia molecules, although there is evidence for other species, including Na^-. At higher metal concentrations, the properties change. The colour becomes a metallic bronze, and reflectivity measurements show that this is associated with a plasma frequency, as explained in Section 2.4.3. The conductivity rises, and at the same time the behaviour of the magnetic susceptibility changes. At low densities the susceptibility decreases with temperature, displaying the Curie–Weiss type of behaviour characteristic of localized unpaired electrons. At higher densities, it becomes almost independent of temperature, which is characteristic of a metal. These changes are illustrated in Fig. 5.21 (b). The observations are consistent with a transition to a metallic state, at around 15 mole per cent in the Na–NH_3 system.

Metal–insulator transitions can be seen in a wide variety of solids. We shall discuss them again from different points of view in the following chapters.

Further reading

More detailed accounts of the Hubbard model tend to be highly theoretical and rather inaccessible to chemists. An introduction to the (mostly physics-based) literature in this area can be found in:

N. F. Mott (1974). *Metal–insulator transitions.* Taylor and Francis.

An extensive account of the properties of transition-metal oxides, including discussion of theoretical models involved, is given by:

J. B. Goodenough (1971). *Prog. Solid State Chem.* **5** 143.

The following papers describe the application of photoelectron spectroscopy to some particular series of solids:

J. K. Lang, Y. Baer, and P. A. Cox (1981). *J. Phys. F: Metal Phys.* **11** 121.
P. A. Cox, R. G. Egdell, J. B. Goodenough, A. Hamnett, and C. C. Naish (1983). *J. Phys. C: Solid State Phys.* **16** 6221.

The ligand field theory forming the background to the discussion of the properties of localized electrons is described by:

B. N. Figgis (1966). *Introduction to ligand fields.* John Wiley and Sons.

Accounts of the magnetic properties of solids are given in many solid state physics books, and in:

J. B. Goodenough (1963). *Magnetism and the chemical bond.* John Wiley and Sons.

The following reviews discuss mixed configuration compounds of lanthanides:

C. M. Varma (1976). *Rev. Mod. Phys.* **48** 218.
M. Campagna, G. K. Wertheim, and E. Bucher (1976). *Structure and bonding* **30** 99.
J. A. Wilson (1977). *Structure and bonding* **32** 57.

The problems involved in applying similar ideas to solid compounds of the actinide elements are discussed in:

J. R. Naegele, J. Ghijsen, and L. Manes (1985). *Structure and bonding* **59** and **60** 198.
M. S. S. Brooks (1985). *Structure and bonding* **59** and **60** 264.

The following accounts describe the polarization catastrophe model of metals and insulators:

P. P. Edwards and M. Sienko (1982). *Acc. Chem. Res.* **15**, 87; *Int. Rev. Phys. Chem.*, **3** 83.

6
Lattice distortions

Band theory is concerned with the behaviour of electrons in a periodic solid. In some ways, this approach is like putting the cart before the horse, because the arrangement of atoms in a crystal is itself determined by the bonding interactions of the electrons. Although the existence of regular crystal structures shows that the bonding forces often produce a periodic lattice, this is not always the case. It is helpful to think of the Jahn–Teller effect discussed in Section 5.3.3. Some electron configurations produce forces that distort the lattice from its ideal regular configuration. The Jahn–Teller effect is strictly applicable only to localized electrons, but there are analogous effects that can occur with electrons in bands. The presence of electrons may sometimes cause a distortion that destroys the simple periodicity of the lattice. This in turn disrupts the band structure, and may have an important influence on the electronic properties of the solid.

Distortions caused by electrons are important in two interesting areas of solid-state chemistry: 'low-dimensional' solids, where the principal electronic interaction is along a chain of atoms, or within a plane; and mixed-valency compounds, where an element is present in two different oxidation states. A breakdown of the crystal periodicity can also happen in other situations, for example when the solid has defects or foreign impurity atoms. The effects of these are discussed in Chapter 7.

6.1 Low-dimensional solids

In Chapter 4 models of one- and two-dimensional solids were used to introduce the concepts of band theory. In fact, many real solids with a one-dimensional electronic structure are now known, but their properties are much more complicated than simple band theory would suggest. The study of such compounds is an active research field, where many important developments have been made by teams of chemists and physicists working together.

6.1.1 Periodic lattice distortions and charge density waves

One of the best-known one-dimensional conductors is the partially oxidized platinum chain compound, $K_2Pt(CN)_4Br_{0.3} \cdot 3H_2O$, which for simplicity is

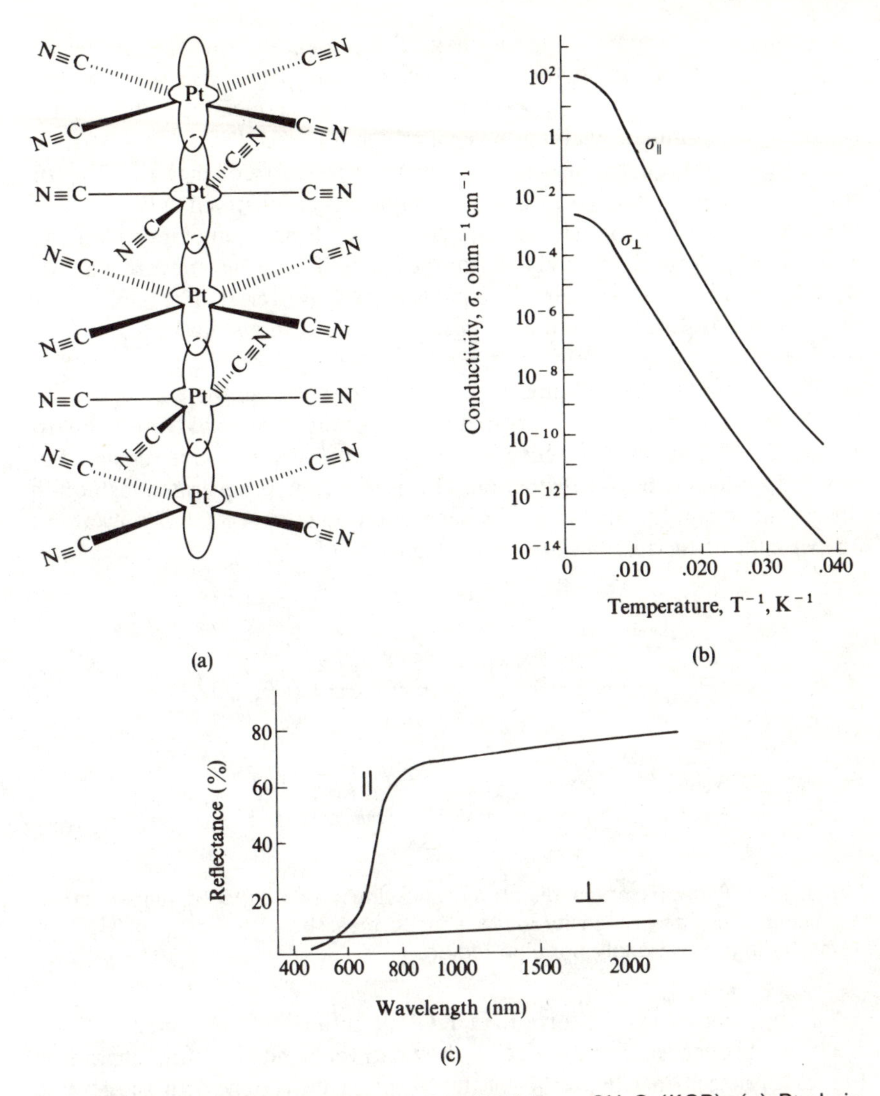

Fig. 6.1 Structure and properties of $K_2Pt(CN)_4Br_{0.3}\cdot 3H_2O$ (KCP). (a) Pt chain structure, showing overlap of d_{z^2} orbitals. (b) Temperature dependence of conductivity parallel and perpendicular chain direction. (c) Optical reflectance in two directions. (After J. S. Miller and A. J. Epstein, *Prog. Inorg. Chem.*, 20 (1976), 1; H. R. Zeller and A. Beck, *J. Phys. Chem. Solids*, 35 (1974), 77; H. P. Geserich *et al.*, *Phys. Stat. Solidi* (A), **9** (1972), 187.)

usually known as KCP. Part of the structure is shown in Fig. 6.1, together with some of the observed properties. The metal atoms form chains along the *c*-axis of the crystal, and the top-occupied band is composed principally of platinum

$5d_{z^2}$ orbitals overlapping along the chain. The presence of the Br^- ions leaves 0.3 holes per platinum in this band, which would otherwise be filled (see Section 3.4.3). At room temperature there is good metallic conductivity, occurring predominantly along the chain directions (see Fig. 6.1(b)). There is also metallic reflectivity, which drops at the plasma frequency of 17×10^3 cm^{-1} (2.1eV); as shown in the Fig. 6.1, the reflectivity is observed only with light polarized parallel to the chains. Crystals of KCP are quite transparent to visible light polarized perpendicular to the c-axis. Such properties are what we would expect for a metal where the band is formed by overlap of orbitals in one dimension only. But KCP is not a true metal, since as shown in the Fig. 6.1, the conductivity declines sharply at temperatures below 150 K. There is a band gap at lower temperatures, although it seems to disappear as the temperature is raised. Diffraction studies also reveal an interesting structural feature: below about 250 K, the spacing between atoms in each chain is not regular, but displays a **periodic lattice distortion**. This is shown in exaggerated fashion in Fig. 6.2(a): in fact the maximum displacement observed in KCP is only about 0.5 per cent of the regular lattice spacing.

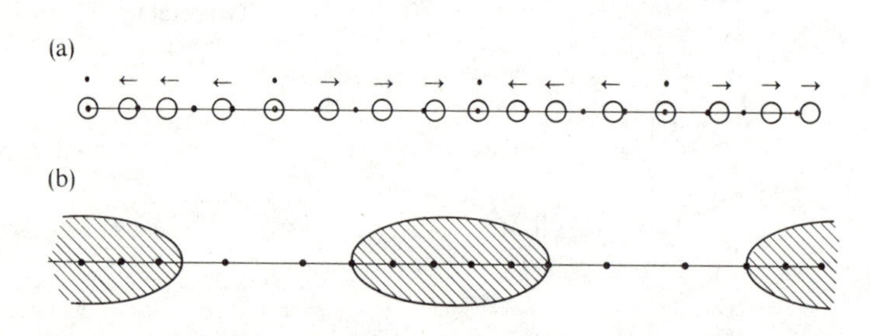

Fig. 6.2(a) Periodic lattice distortion, showing modulation of regular chain spacing. (b) Charge density wave: shaded area shows build-up of electron density in more strongly bonding region.

The periodic lattice distortion in KCP is intimately connected with its electronic properties, and its effect is shown in the band structure diagram in Fig. 6.3. As seen in Chapter 4, electrons interact strongly with the atomic potential when their wavelength is two lattice spacings, and an energy gap is produced in the one-dimensional band structure. For the normal lattice spacing c, the appropriate wavelength is $2c$, corresponding to the wave-number π/c at the top of the band. The periodic distortion, however, gives rise to a 'superlattice' with a longer period, and the interaction of electron waves with this superlattice period produces another gap in the diagram. The diffraction studies show that the superlattice spacing in KCP is 6.6 times the regular lattice spacing, and this is just the value required to give a gap at the top-filled level in the band. Thus KCP is a semiconductor at low temperatures. At higher

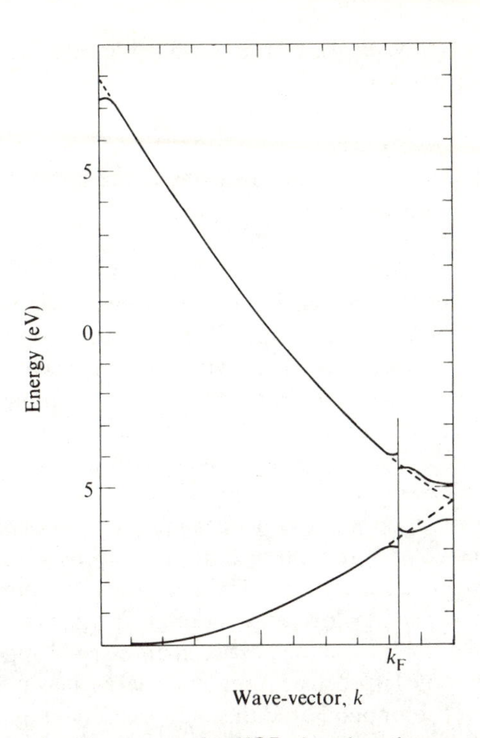

Fig. 6.3 Calculated band structure for KCP showing the top occupied and the bottom empty bands, and the gap produced by the periodic lattice distortion. k_F is the wave-vector for the top-filled level in the lower band. (After H. R. Zeller and P. Bruesch, *Phys. Stat. Solidi* (B), **65** (1974), 537.)

temperatures, the periodic distortion is smeared out by the thermal vibrations of atoms, and the gap disappears.

Another consequence of the periodic distortion is shown in Fig. 6.2 (b). The bonding between atoms is stronger where they are more closely spaced, and the electrons have lower energy in these regions. They therefore tend to concentrate at intervals along the chain. This periodic build-up of electron density is known as a **charge density wave.** The gap produced by the periodic distortion is essentially the energy required to move an electron from a more strongly bound region, to one where the atoms are further apart.

The occurrence of periodic lattice distortions and charge density waves is characteristic of one-dimensional conductors, and the case of KCP is an illustration of the **Peierls theorem**, which asserts that a true one-dimensional metal does not exist. It was seen in Chapter 4 that any periodic potential gives rise to a band gap in one dimension. If the gap is just at the top-filled level, it will give rise to a net lowering of the energy of occupied orbitals, and hence of the total energy of the solid. The superlattice period required to produce such a gap is just half the wavelength of the top-filled level, which is related to its wave-vector by equation 4.14: (see p. 82)

$$\lambda = 2\pi/k$$

Thus is if k_F is the wave-vector of electrons at the Fermi level, the periodic distortion will have a wavelength:

$$\lambda_D = \pi/k_F. \tag{6.1}$$

Properties similar to those of KCP have been observed in many other compounds containing chains of metal atoms with a square-planar ligand coordination. This geometry is found with elements such as iridium, nickel, and palladium, as well as with platinum. Metal chain compounds show some of the best examples of one-dimensional electronic structure, since the chains are well separated from each other by the ligand groups, and the three-dimensional electronic interactions are very small.

The Peierls theorem does not apply generally to two or three dimensions, since as seen in Chapter 4, the interaction of electrons with a weak periodic potential does not give an energy gap. There is some distortion of the density of states curves, but any energy lowering of the electrons is not usually sufficient to overcome the elastic interactions between atoms, which generally favour a regular lattice. Periodic lattice distortions and charge density waves *are* known in a number of layer compounds, but this only seems to happen when the band structure has rather special features. The best-known example is tantalum disulphide (TaS_2), where the effect is much more pronounced than in KCP. At low temperatures the periodic distortion in TaS_2 involves an atomic displacement of 25 pm, and a charge displacement equivalent to nearly one electron per tantalum. As in KCP, the period of the displacement is related to the band filling, and is altered by changing the number of metallic electrons. This can be done, for example, by replacing some Ta^{4+} (d^1) with Ti^{4+} (d^0). Unlike KCP, TaS_2 is still metallic below the temperature at which the distortion occurs. This is because an energy gap is only produced for electrons moving in the direction corresponding to the distortion. In two dimensions, electrons have the freedom to move in other directions, and so to avoid the gap.

Periodic lattice distortions are often **incommensurate,** and as in KCP they do not correspond to an integral number of normal lattice spacings. The distortion in TaS_2 is incommensurate between 350 K, when it first appears, and 200 K. Below 200 K, however, it changes its nature, and becomes **commensurate** and locked onto the lattice. Incommensurate modulations of the structure are known in a number of non-metallic compounds, such as sodium nitrite ($NaNO_2$). In these solids, however, the periodic modulation is not caused by electronic effects, but by a subtle combination of ionic and elastic interactions. Such incommensurate structures are not well understood, and are the subject of current research.

6.1.2 Half-filled bands

When the band in the regular structure is just half full, the Peierls distortion has a particularly simple form. The periodic distortion predicted by equation 6.1 now corresponds to exactly two lattice spacings, and gives an alternation of bond lengths along the chain of atoms. This can be seen in Fig. 6.4, where the splitting of the band due to the doubling of the lattice periodicity is also shown. The concentration of electron density can be interpreted as a simple bond alternation along the chain, with the electrons trapped in bonding orbitals between the closely-spaced pairs of atoms. The band gap produced by the distortion in this case is the energy difference between these bonding orbitals, and the corresponding antibonding ones.

The best-known example of a Peierls distortion in a half-filled band is provided by polyacetylene, $(CH)_x$, which can be prepared by polymerizing ethyne (acetylene), C_2H_2. The properties of polyacetylene have been extensively investigated in recent years, and some aspects are still controversial. The simplest band picture of π orbitals overlapping in a regular chain would be just

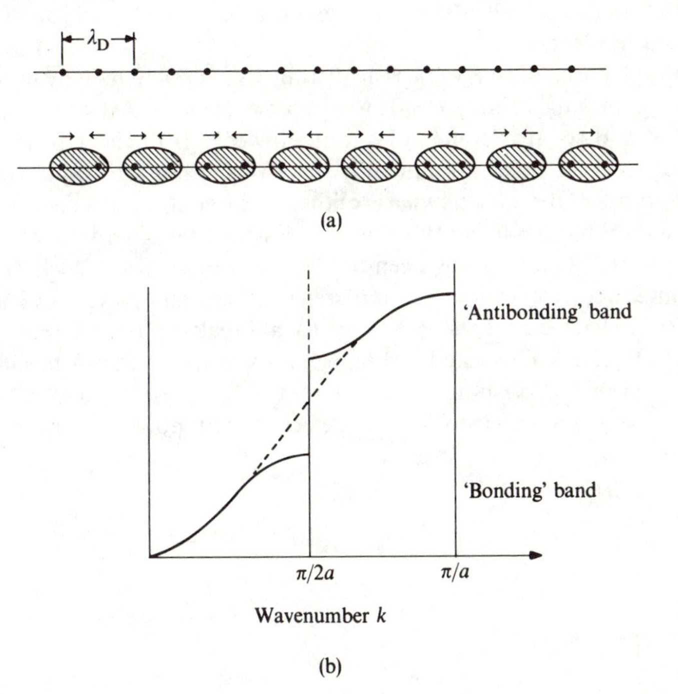

Fig. 6.4 Peierls distortion in a half-filled band. (a) Distortion period of two lattice spacings gives an alternation of bond lengths, with electrons concentrating in the more bonding regions. (b) Band structure diagram, showing the splitting into a 'bonding' and an 'antibonding' band.

like the one-dimensional s band of Section 4.1.2. The simplest model of polyacetylene should have a half-filled π band, and be metallic. Longuet-Higgins and Salem showed however, that a bond alternation should give a lowering of energy in the LCAO theory, just as in the free-electron model. Because the structure of polyacetylene is rather disordered, it is difficult to determine accurately. It appears however that there is an alternation in the C–C bond lengths, of about 6 pm. This is much smaller than the difference between normal single and double bonds, and although the structure is normally drawn with alternating single and double bonds this exaggerates the

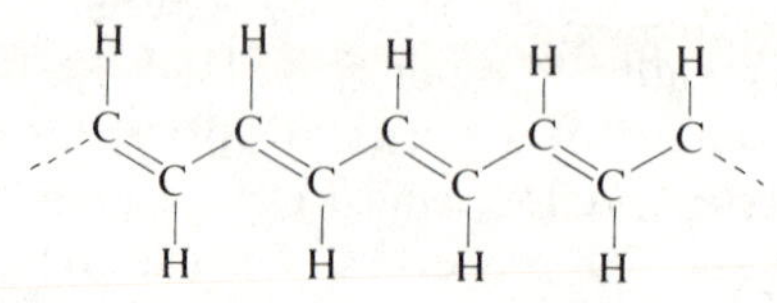

effect. Polyacetylene is non-metallic, with a band gap of about 2 eV. Part of the gap at least must come from the bond alternation, but there may also be a contribution from electron repulsion, giving a Mott–Hubbard gap of the kind described in Chapter 5.

Although pure polyacetylene is insulating, its conductivity may be increased greatly by doping it with electron acceptors such as AsF_5. These remove electrons, and so leave some holes in the lower part of the π band. There is evidence that the holes are not fully delocalized, but are trapped at rather unusual types of defects known as **solitons.** (The concept of a soliton arises in mathematical physics from the solutions found for certain non-linear differential equations. The idea has been applied in diverse areas, such as domain boundaries and defects in solids, and even to elementary particles.) The soliton in polyacetylene corresponds simply to a break in the pattern of bond alternation, and is illustrated in Fig. 6.5. As shown, there can be charged or neutral solitons, corresponding either to a missing π electron, or to a non-bonding unpaired electron at the defect. When polyacetylene undergoes

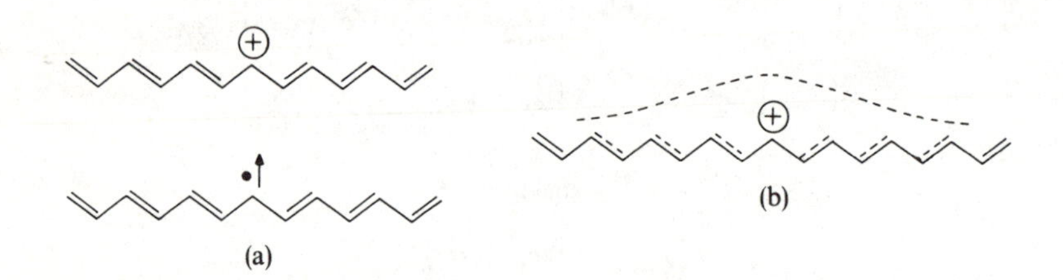

Fig. 6.5 Solitons in polyacetylene. (a) Charged and neutral solitons, showing the break in the pattern of bond alternation. (b) More realistic picture of the charged soliton, showing the spread of charge over several atoms, and the gradual change of bond alternation.

isomerization from the *cis* form to the thermodynamically more stable *trans* modification, it appears that 'mistakes' occur in the bond alternation, leaving neutral solitons which can be detected from the ESR signal of the unpaired electrons. On doping, the conductivity rises, simultaneously with a fall in the ESR signal. This is because the non-bonding unpaired electrons are the easiest to ionize. It is the *inverse* relation between unpaired electrons and conductivity that forms the strongest evidence for the soliton idea. If conductivity occurred by delocalization of holes in a band, then the ESR should show a *positive* correlation with conductivity, since each hole created would leave behind an unpaired electron.

More detailed calculations suggest that the charged soliton is probably not as localized as the picture in Fig. 6.5(a) suggests. The positive charge may be spread out over as many as 15 atoms, and the bond alternation changes gradually in this region, as shown in Fig. 6.5(b).

Another solid where non-metallic behaviour is associated with a bond alternation is the transition-metal compound vanadium dioxide, VO_2. Above 340 K it has the rutile structure of TiO_2, and is metallic. Although the structure is not obviously low-dimensional, the closest approach of vanadium atoms occurs along chains parallel to the *c*-axis, where the metal–oxygen octahedra share edges (see Fig. 6.6). It appears that the most important electronic interactions between the vanadium 3*d* orbitals occur along these chains. VO_2 has one electron in the vanadium 3*d* band, and this gives metallic conductivity in the high-temperature form. However, below 340 K the solid becomes non-metallic, and undergoes a structural distortion that has the effect of producing an alternation in the V–Vdistances along the chain. The electronic consequences of the distortion are shown in Fig. 6.7. In the regular structure, the electron occupies the lower part of a band coming from overlap of the three t_{2g}-like *d* orbitals. In the distorted structure, however, the pairing of the vanadium atoms splits off a band, composed of *d* orbitals that are bonding between the closely-spaced pairs of metal atoms. This is rather like polyacetylene, and in a similar way, the band can only accommodate two electrons per pair of vanadium atoms. Thus the distorted structure leads to a band gap in VO_2. Figure 6.8(a) shows photoelectron spectra of the band above and below the transition temperature. The sharp Fermi edge expected from the top-occupied level in a metallic conduction band can be seen at higher temperatures, and

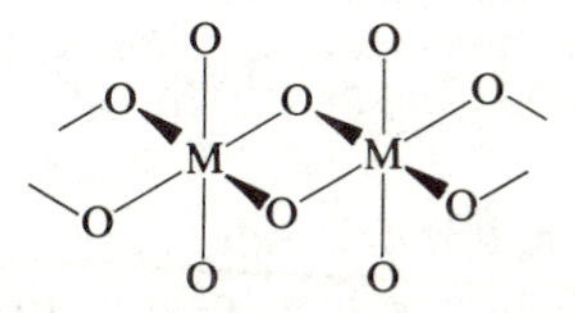

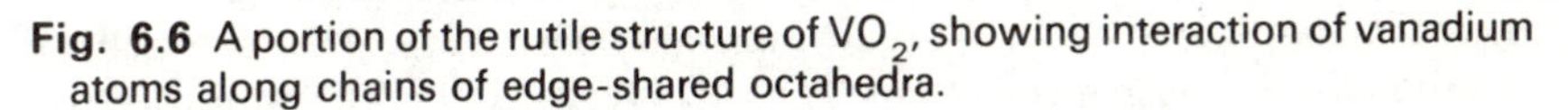

Fig. 6.6 A portion of the rutile structure of VO_2, showing interaction of vanadium atoms along chains of edge-shared octahedra.

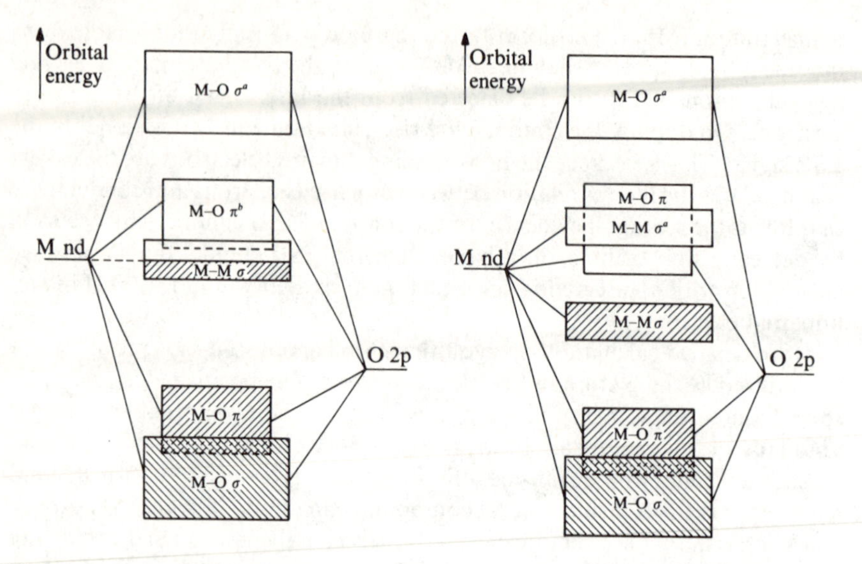

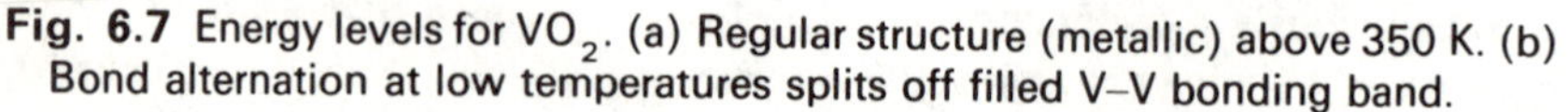

Fig. 6.7 Energy levels for VO_2. (a) Regular structure (metallic) above 350 K. (b) Bond alternation at low temperatures splits off filled V–V bonding band.

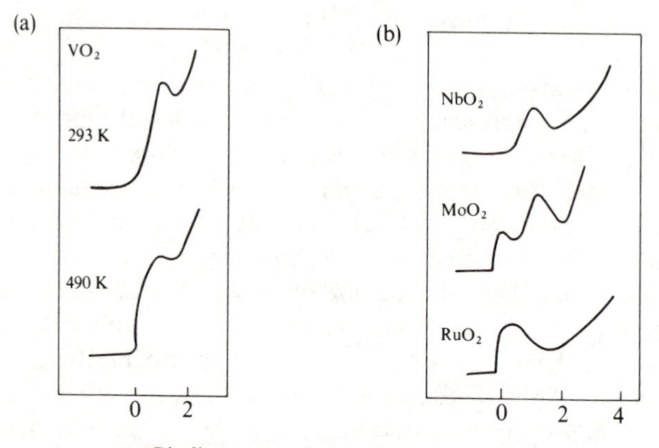

Fig. 6.8 Photoelectron spectra of some MO_2 compounds. (a) VO_2 above and below transition temperature, showing Fermi edge in the metallic form. (b) 4*d* dioxides. Note the split *d* band in MoO_2. (After N. Beatham and A. F. Orchard, *J. Electr. Spectr. Relat. Phenom.*, **26** (1979), 77.)

disappears at the transition. It is interesting that the $(4d)^2$ dioxide MoO_2 also has a distorted rutile structure similar to VO_2. In this compound however, there is an additional electron occupying higher parts of the *d* band, which are non-bonding between metal atoms. Thus MoO_2 is metallic. The photoelectron spectra of a number of $4d$ metal dioxides are shown in Fig. 6.8(b). NbO_2 $(4d)^1$ is like VO_2, being distorted and non-metallic, with the lower part of the d band filled. The spectrum of MoO_2 shows two $4d$ peaks, the one at higher binding energy corresponding to the split-off metal–metal bonding band. RuO_2, with $(4d)^4$ has the regular rutile structure, because the additional $4d$ electrons would now have to occupy M–M antibonding levels in the distorted form, which would be energetically unfavourable. Like MoO_2 it is metallic, but no splitting of the $4d$ band can be seen in the spectrum. The sharp Fermi edge in the spectra of the metallic oxides is also noticeable, as compared with non-metallic NbO_2.

6.1.3 Molecular metals

In previous chapters molecular solids have been dismissed somewhat, as having rather uninteresting electronic properties essentially similar to those of individual molecules. In fact, there is an interesting class of molecular compounds where rather stronger inter-molecular interactions lead to high conductivity and other remarkable properties. These are sometimes called **molecular metals.**

The first-discovered molecular metal was TTF:TCNQ, a 1:1 solid compound of tetrathiafulvalene (TTF) and tetracyanoquinodimethane (TCNQ). The structures of the molecules and the solid are shown in Fig. 6.9. The solid structure of TTF:TCNQ is rather unusual, having alternate stacks each composed entirely of one type of molecule. The electrical conductivity of TTF:TCNQ is shown as a function of temperature in Fig. 6.10. There is a maximum at around 80 K, with an apparent activation energy at lower temperatures, very similar to the case of KCP (Fig. 6.1); as with KCP diffraction studies show that the energy gap is associated with a periodic distortion in the spacing of molecules in the stacks.

The properties of TTF:TCNQ can only be explained by assuming some degree of charge transfer between TTF and TCNQ. TCNQ is a good electron

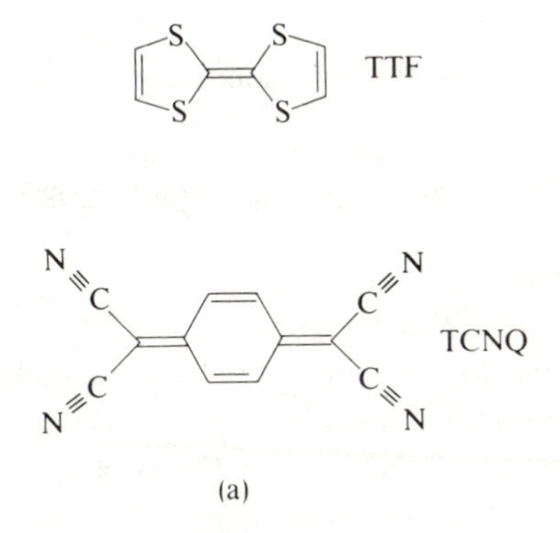

Fig. 6.9(a)

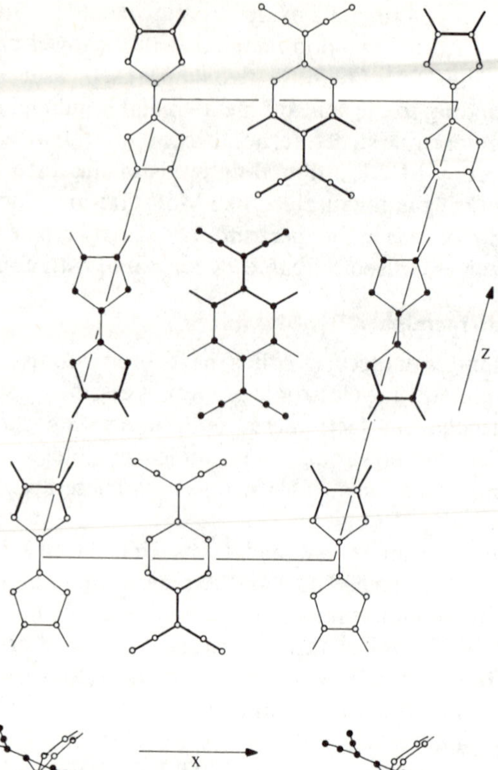

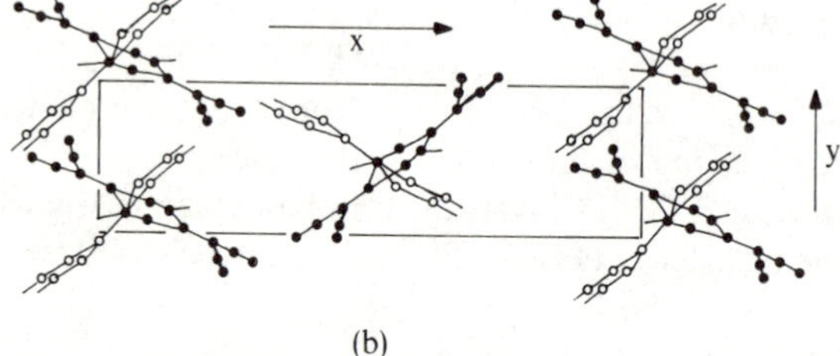

(b)

Fig. 6.9 TTF: TCNQ. Stuctures of (a) component molecules tetrathiafulvalene (TTF) and tetracyanoquinodimethane (TCNQ), and (b) solid TTF: TCNQ showing alternate stacks of segregated TTF and TCNQ molecules. (After D. Jerome and H. J. Schulz, *Adv. Phys.*, **31** (1982), 299.)

acceptor, and can form ionic salts such as $K^+(TCNQ)^-$. On the other hand, electrons are fairly easily removed from the top-filled orbitals in TTF, which are largely sulphur lone-pair in character. Thus the conductivity comes from two partially filled bands, composed of the top-filled orbital in TTF, and the bottom empty orbital in TCNQ. Since the overlap between molecules is much greater within a chain than between chains, the band structure is one-dimensional in character, and there is a Peierls distortion as in KCP. As shown

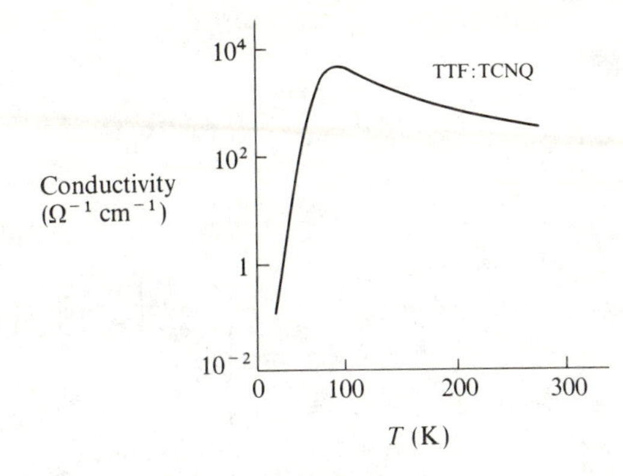

Fig. 6.10 Conductivity of TTF: TCNQ as a function of temperature. (After J. B. Torrance, *Acc. Chem. Res.*, **12** (1979), 79.)

in Section 6.1.1, the period of the lattice distortion can be related to the number of electrons in the band. Structural studies on TTF:TCNQ show that the charge distribution corresponds to a transfer of 0.69 electron per molecule from TTF to TCNQ.

Some compounds with properties similar to TTF:TCNQ are known, and a systematic study suggests that the following factors are required for high conductivity:

1. Segregated stacks of 'anions' and 'cations' as in the structure of Fig. 6.9.
2. Partial charge transfer ($0 < q < 1$) between molecules.

In the segregated stack structure of TTF:TCNQ, the predominant overlap is between molecules of one type, with orbitals at the same energy. This is more efficient in forming a band than is overlap between different molecules. Compounds with mixed stacks, where anions and cations alternate, also tend to have a full transfer of one electron per molecule, and this is also unfavourable to conductivity. Although the anion and cation bands are now formally half full, the width of these bands is quite small, less than 1 eV. Thus with integral charge transfer the electrons are localized by repulsion effects, and there is a Mott–Hubbard splitting as discussed in Chapter 5.

The importance of partial charge transfer is illustrated in a series of TCNQ salts with partners of different electron donor ability. Figure 6.11 shows how the conductivity of the 1:1 solids varies with the electrode potential of the partner. When the donor E^0 is around $+1$ V no charge transfer takes place, as it is too difficult to remove electrons from the filled orbitals. These solids are non-conducting because there are no partially occupied bands. With highly reducing donors having E^0 less than 0, the charge transfer is integral, and electrons are localized by repulsion in the narrow bands. The highly

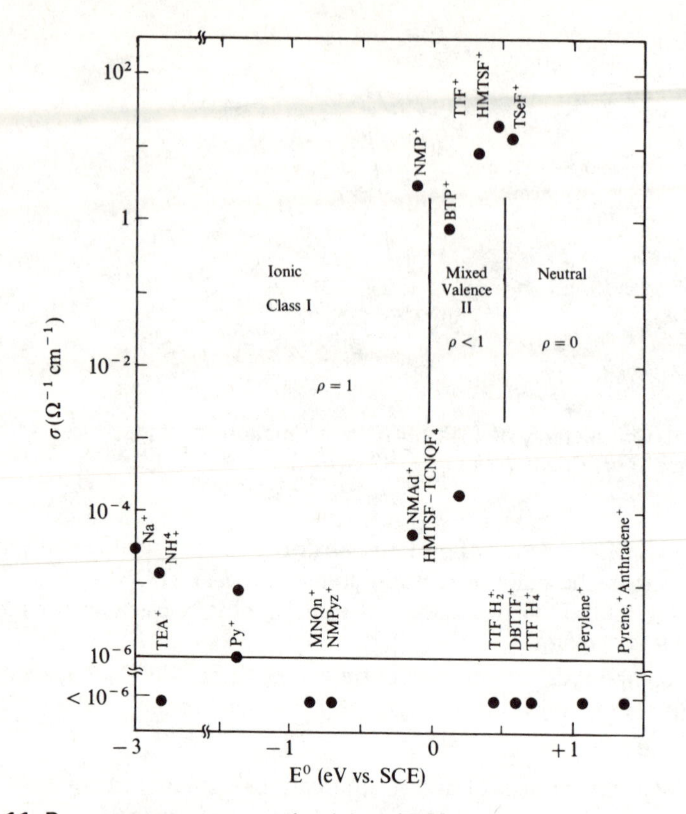

Fig. 6.11 Room-temperature conductivity of TCNQ salts as a function of cation E^0. (After J. B. Torrance, *Acc. Chem. Res.*, **12** (1979), 79.)

conducting compounds like TTF:TCNQ occur in the range $0 < E^0 < 0.5$, where the partial charge transfer may be deduced from the occurrence of periodic lattice distortions. The detailed reason for such a partial ionic character is not fully understood, but it seems to involve a balance of the Coulomb forces between ions, and bonding in the anion and cation bands.

Another series of molecular metals is that formed by salts of TMTSF, tetramethyl-tetraselenofulvalene (shown below), with inorganic anions.

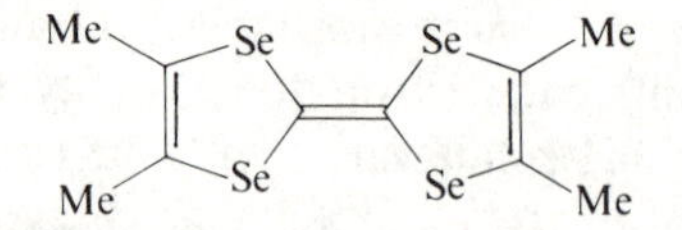

Compounds such as $(TMTSF)_2^+\,(ClO_4)^-$ and $(TMTSF)_2^+\,(PF_6)^-$ have stacks of adjacent TMTSF molecules, with 'partially' ionic $TMTSF^{0.5+}$. Thus they fullfil the criteria just discussed for high conductivity. These solids differ from TTF:TCNQ however, in that their conductivity does *not* fall off at low temperatures; indeed at very low temperatures they become superconducting. It seems that one reason for the difference is that there is significant overlap *between* the TMTSF stacks. Thus the electronic structure is not so one-dimensional. Similar behaviour is shown by the inorganic polymer $(SN)_x$. The chain structure drawn with the normal valencies shows that there must be an unpaired electron associated with each SN unit. Thus there is a partially filled

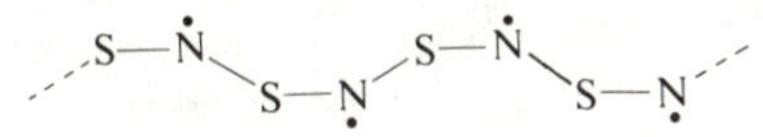

band, which gives rise to metallic conductivity. However, $(SN)_x$ does not show one-dimensional behaviour, and like the TMTSF salts, is superconducting at low temperatures. Once again, there are significant interactions between the chains, which suppress the Peierls distortion expected in a purely one-dimensional conductor.

6.2 Polarons

The low-dimensional phenomena that have just been discussed can be described as **collective electron** effects, as they are the result, not of individual electrons, but of a partially filled band. In ionic solids, single electrons or holes can also produce more localized lattice distortions, resulting from their electrostatic interaction with neighbouring ions. This type of distortion accompanies the electron as it moves through the lattice, and the result is known as a **polaron**. When the distortion is sufficiently strong, an electron or hole may be trapped at a particular lattice site, and can only move by thermally activated **hopping** through the solid.

6.2.1 Small and large polarons

Let us consider the effect of putting an electron in an unfilled orbital on a particular atom in a solid. The overlap of orbitals on adjacent atoms gives rise to a band. The overlap effect will therefore tend to delocalize the electron, and its energy will be decreased by an amount $W/2$, where W is the total bandwidth (see Fig. 6.12). On the other hand, if we force the electron to stay on one atom, the charge will polarize surrounding atoms, which will also lower the energy. The approximate formula for the polarization energy given previously (see equation 2.2 on p. 29) is:

$$\Delta E = -e^2/(8\pi\varepsilon_0 r)(1-1/\varepsilon_r) \tag{6.2}$$

where r is the radius of the orbital, and ε_r the relative dielectric constant of the solid. It is convenient to separate the polarization into two parts. The first is that due purely to the electron clouds. This is the effect that contributes to the high-frequency or optical dielectric constant, ε_{opt}. Thus the energy lowering

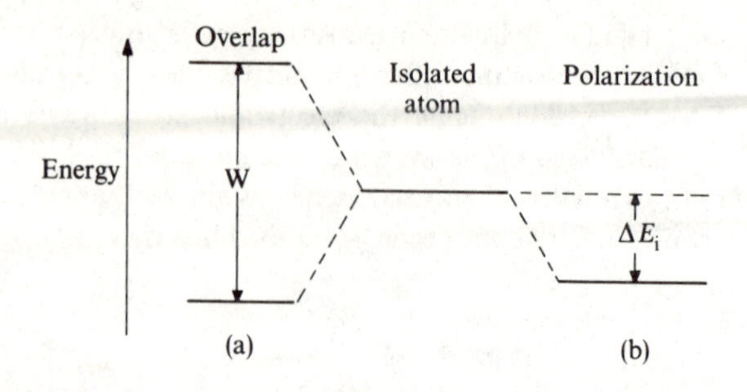

Fig. 6.12 Electron in a solid can lower its energy either through delocalization in a band (a), or by polarizing the surrounding lattice (b).

from electronic polarization is:

$$\Delta E_e = -e^2/(8\pi\varepsilon_0 r)(1 - 1/\varepsilon_{\mathrm{opt}}) \tag{6.3}$$

The other part is that due to the displacement of ions from their regular lattice positions. The static dielectric constant ε_s contains contributions from both ionic and electronic effects. Putting ε_s in equation 6.2 would give the total polarization energy, and so the ionic contribution alone must be the difference between this and the electronic part (equation 6.3). That is:

$$\Delta E_i = -e^2/(8\pi\varepsilon_0 r)(1/\varepsilon_{\mathrm{opt}} - 1/\varepsilon_s) \tag{6.4}$$

If the electron is now able to move in the band formed by the orbital overlap, the electronic polarization ΔE_e can follow it, but the part ΔE_{i} cannot, since this term depends on the relatively slow motions of atoms as a whole. Thus equation 6.4 gives the polarization energy that is *lost* if the electron is delocalized. The actual behaviour of the electron will depend on whether it achieves a lower energy by remaining on one atom, and getting the extra polarization ΔE_{i}, or by delocalizing, and having the associated stabilization $W/2$. If:

$$W/2 < |\Delta E_{\mathrm{i}}| \tag{6.5}$$

it will be the localized state that is more stable, and the electron will be effectively trapped by the local distortion that it creates round a given lattice site. This is called a **small polaron**, and is illustrated in Fig. 6.13 for an electron in a metal orbital in an oxide. The same thing can happen for holes in a filled band. The formation of a small polaron has a simple chemical interpretation. Introducing an extra electron or hole changes the oxidation state of one atom in the solid. This changes the ionic radius, which produces the local distortion that can trap the charge. Equation 6.5 expresses the condition for valencies to be trapped by the difference of ionic radii. If this criterion is not satisfied, the electrons or holes will be delocalized, and all ions will have the same fractional

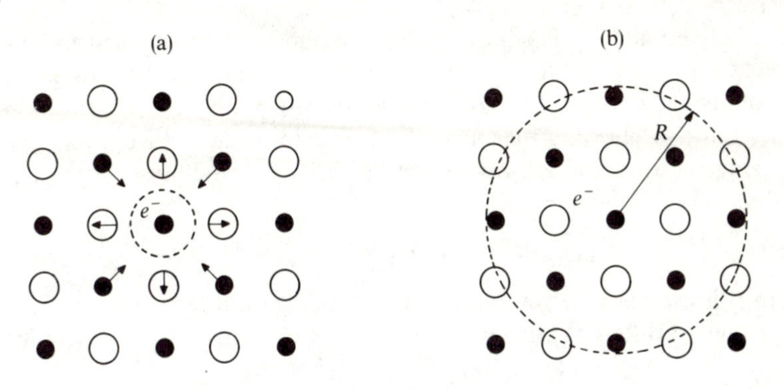

Fig. 6.13(a) Small polaron, showing the distortion of the lattice round an electron trapped at a metal ion; (b) electron distribution of a large polaron of radius R, formed in a metal oxide MO. ●, Metal ion; ○, O^{2-}.

oxidation state. As will be seen later, these different possibilities are important in understanding the electronic properties of mixed valency compounds.

It must be emphasized that equation 6.4 gives only a very approximate estimate of the ionic relaxation energy, as our argument was based on a crude electrostatic picture. The model nevertheless gives a useful qualitative guide to where small polarons are likely to be found. For a large value of ΔE_i predicted by equation 6.4, we require orbitals of small radius, in a solid where ε_{opt} is small, and ε_s is large. As discussed previously in Section 3.2.3, there is likely to be a significant difference in the two dielectric constants in a compound with a fair degree of ionic character. In a purely covalent solid, such as Si, there is no ionic contribution to the polarization, and ε_{opt} and ε_s are the same. Even in ionic compounds, values of ΔE_i are quite small, around 1 eV, and so we need to have narrow bands to satisfy the condition (equation 6.5) for localization. Holes in the narrow valence bands in halides are expected to give small polarons, but not electrons in the broader conduction bands. In fact, holes are trapped in alkali and silver halides, but in a more complicated way than the simple small polaron picture would suggest. The small polaron picture is most commonly applied to transition-metal compounds, especially in the 3d series where the d bands are rather narrow.

The small polaron represents an extreme case of electron–lattice interaction. Some degree of polarization occurs in ionic compounds even when the condition for complete trapping is not satisfied. To see how this happens, imagine an electron in a conduction band, localized in a region of radius R, which may be many lattice spacings (Fig. 6.13(b)). To localize the electron requires an increase of kinetic energy, given roughly by the ground state of the electron-in-a-box model:

$$E_{\text{kin}} = h^2/(2m^* R^2) \tag{6.6}$$

The effective mass m^* has been written instead of the free-electron mass, to correct for the fact that the electron is moving in a band, and not in free space. As explained in Chapter 4, narrow bands give rise to high effective mass. E_{kin} is always positive, showing that there is nothing to be gained from this term by localizing the electron. However, there is also the ionic polarization term to consider. From equation 6.4:

$$\Delta E_i = -e^2/(8\pi\varepsilon_0 R)(1/\varepsilon_{opt} - 1/\varepsilon_s). \tag{6.7}$$

Adding these together gives the total energy of the localized state of radius R. It can be seen by differentiating with respect to R, that there is always a minimum where:

$$R = \frac{8\pi\varepsilon_0 h^2}{e^2 m^*(1/\varepsilon_{opt} - 1/\varepsilon_s)} \tag{6.8}$$

As before, it appears that localization is favoured by narrow bands (large m^*), and by a large difference between ε_{opt} and ε_s. However, this equation shows that some degree of localization of the electron by ionic polarization is *always* favourable in a compound with ionic character. When the radius R is more than the lattice spacing, this is called a **large polaron**.

The argument just presented suggests that an electron should be trapped by the lattice even when a large polaron is formed. This is not correct however, and a fuller analysis must take account of the way in which the polarization is able to follow the electron as it moves. The speed at which ions can move is given by their vibrational frequency ω_v: for electrostatic polarization effects, the **optical phonon frequency** should be used, corresponding dipole-active vibrations where ions of different charge move out of phase. The detailed theory of moving polarons is quite difficult, but the result is that the tendency to form large polarons depends on the **Fröhlich polaron coupling constant**:

$$\alpha_p = e^2/(4\pi\varepsilon_0\hbar)(m^*/2\hbar\omega_v)^{1/2}(1/\varepsilon_{opt} - 1/\varepsilon_s) \tag{6.9}$$

Typical values of α_p, calculated for electrons in conduction bands, are: KCl, 3.7; GaAs, 0.03; $SrTiO_3$, 4.5. Large values correspond to strong polaron formation, and as might be expected, this happens in the more ionic solids. Since the polarization moves with the electron, the effective mass is increased, and the mobility of electrons may be considerably reduced. With large polarons, the mobility declines rapidly with increasing temperature, as more vibrations become thermally excited. This is quite different from the thermally activated mobility found for small polarons, which is described in the next section.

6.2.2 Properties of small polarons

The formation of a small polaron, when electrons or holes are trapped at particular lattice sites, has an important influence on the electronic properties of a solid. Let us consider how an electron can move from its trapped position to neighbouring lattice sites. It is simplest to think of just two atoms, rather than a complete lattice, with an electron initially on one of them. This is illustrated in Fig. 6.14, and can be imagined as a small part of the lattice of a transition-metal oxide, with one electron in the conduction band. The trapping is associated with an increase of the metal–oxygen bond distances, as shown. If the electron moves to the other site, the distortion will follow it. Figure 6.14 shows potential curves for the two states, drawn against a vibrational coordinate that represents the distortion. This picture is called the **configuration coordinate** model. The middle of the diagram corresponds to the symmetrical configuration, where the electron has the same energy on both atoms, so that the curves cross. In fact, around this point the overlap of the orbitals on each atom can give rise to bonding and antibonding combinations, and a splitting of the curves as shown. If the electronic interaction is large enough, the symmetrical point is lowest in energy. This represents the delocalized state, where the electron is not trapped. The simple splitting into two electronic orbitals in the dimer is the same effect that gives the electronic bandwidth of the complete solid. As we have seen above, trapping does not occur if the bandwidth is sufficiently large.

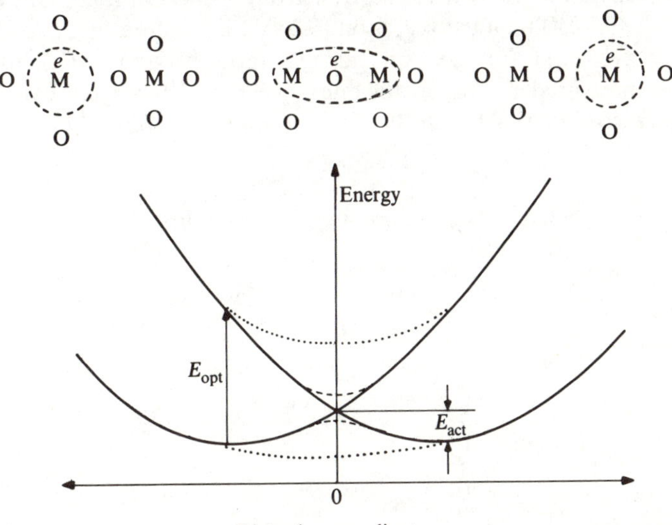

Fig. 6.14 Configuration coordinate diagram for a pair of metal atoms from an oxide MO, with one extra electron. The horizontal axis is a distortion coordinate, with atomic positions shown above. The two energy curves are for electrons on each atom. The effect of weak electronic interaction at the cross-over point is shown (––––): strong interaction (·····) gives a symmetrical delocalized ground state.

We shall consider the case where the electronic interaction is small, so that the lowest energy states for the electron are either on one atom or the other. In order to move between the atoms, the electron must cross a barrier. Thus there is an activation energy for **hopping** of the electron between neighbouring atoms. It is this activated mobility of the electrons which is most characteristic of the small polaron trapping. There will actually be a small probability of the electron **tunnelling** through the barrier. However, since such tunnelling involves the motion of heavy ions, it will normally be very slow, and only significant at very low temperatures.

A good example of hopping conductivity is shown in the compound $Co_{1-x}Fe_{2+x}O_4$. The stoichiometric compound ($x = 0$) may be formulated as $Fe^{3+}[Co^{2+}Fe^{3+}]O_4$, and has the inverse spinel structure, with Fe^{3+} in tetrahedral sites in an oxide lattice, and the octahedral sites shared between Fe^{3+} and Co^{2+}. When there is some excess iron, this replaces Co^{2+}, giving $Fe^{3+}[Co^{2+}_{1-x}Fe^{2+}_{x}Fe^{3+}]O_4$. Electrons may hop between Fe^{2+} and Fe^{3+} in the octahedral sites. On the other hand, the compound may have excess cobalt, which then replaces Fe^{3+}, to give $Fe^{3+}[Co^{2+}Co^{2+}_{x}Fe^{3+}_{1-x}]O_4$. In this case, electron hopping occurs between the Co^{2+} and Co^{3+}. Figure 6.15 shows the activation energy observed for the conductivity, as a function of stoichiometry. It can be seen that there is a much larger activation energy for hopping between cobalt ions (0.5 eV) than between iron (0.2 eV). It is interesting that the same difference is found for electron transfer rates between M^{2+} and M^{3+} complexes measured in solution: the rates for Fe are much faster than those for Co. The reason for this lies in the electron configurations:

$$Fe^{2+}: (t_{2g})^4 (e_g)^2; \quad Fe^{3+}: (t_{2g})^3 (e_g)^2$$

$$Co^{2+}: (t_{2g})^5 (e_g)^2; \quad Co^{3+}: (t_{2g})^6 \quad \text{low spin}$$

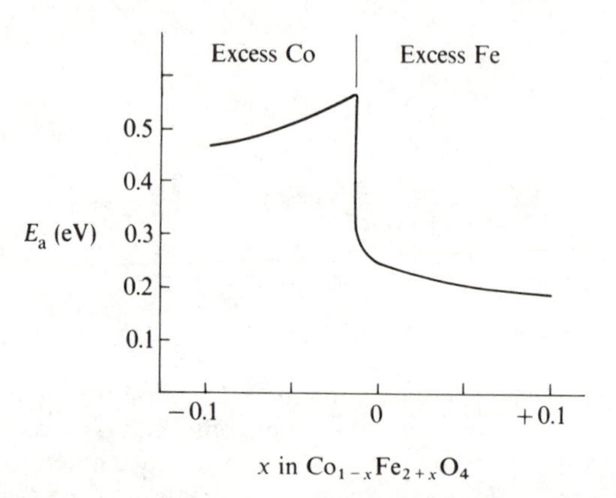

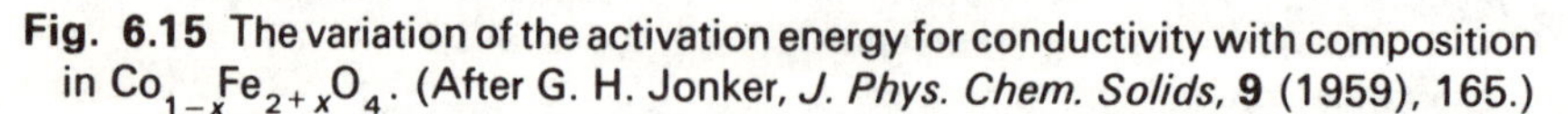
Fig. 6.15 The variation of the activation energy for conductivity with composition in $Co_{1-x}Fe_{2+x}O_4$. (After G. H. Jonker, *J. Phys. Chem. Solids*, **9** (1959), 165.)

Electron transfer between Fe^{2+} and Fe^{3+} involves a t_{2g} electron of relatively non-bonding character. Thus the change in bond length between the two oxidation states is small. It can be seen from the configuration coordinate diagram (Fig. 6.15) that the activation energy for crossing the barrier is proportional to the square of the displacement. Transfer of an electron in the cobalt case requires a much greater electronic change, involving two strongly antibonding e_g electrons, so that the bond length change is larger.

The higher curve on each side of the configuration coordinate diagram is that of the electron on the 'wrong' atom. According to the Franck–Condon principle, a spectroscopic transition of the electron occurs vertically between the potential energy curves, since the heavy atoms do not have time to move appreciably during such a transition. The transition shown in Fig. 6.14 corresponds to an excitation of the electron from its trapped state on one atom, to a neighbouring atom which is not in the correct ground-state geometry to receive it. As will be seen, such **intervalence transitions** are characteristic of mixed valency compounds in which the electrons are trapped by lattice distortions. If the electronic interaction is small, corresponding to the solid lines in the diagram, the thermally activated hopping occurs through the crossing-point of the two parabolic curves. The $y = x^2$ form of a parabola shows that in this case:

$$E_{\text{opt}} = 4E_{\text{act}}. \tag{6.10}$$

This is observed in some small-polaron solids.

Although we have considered isolated electrons or holes, the extra charge that they introduce into the solid must be compensated somehow. Changes in stoichiometry are usually associated with defects or impurities. The electronic properties of the solid are also influenced by the defects and as will be seen in the Chapter 7, there are extra trapping effects, in addition to those caused by polaron formation. In many non-stoichiometric solids, it is difficult to separate the different contributions to the electron trapping.

6.3 Mixed valency compounds

A mixed valency compound may be defined as one in which an element is present in different oxidation states, or in which the formal oxidation state is fractional. Mixed valency has already been mentioned several times, and it is useful here to give a more descriptive account of this interesting class of compounds.

6.3.1 Classification of mixed valency compounds

The interesting properties of mixed valency compounds arise from the possibility of electron transfer between atoms with different oxidation state. The behaviour can vary greatly, according to how easy such a transfer is. This is

Table 6.1
Some mixed valency compounds

Class	*Compound*	*Oxidation states*
Class I	KCr_3O_8	Cr(III, VI)
	$GaCl_2$	Ga(I, III)
Class II	Eu_3S_4	Eu(II, III)
	$Fe_4[Fe(CN)_6]_3.xH_2O$	Fe(II, III)
	Na_xWO_3 ($x < 0.3$)	W(V, VI)
	$[(C_2H_5NH_2)_4PtCl](ClO_4)_2$	Pt(II, IV)
	$CsAuCl_3$	Au(I, III)
	Cs_2SbCl_6	Sb(III, V)
Class IIIA	$Nb_6Cl_{14}{\cdot}8H_2O$	Nb(2.33)
Class IIIB	Na_xWO_3 ($0.3 < x < 0.9$)	W($6-x$)
	Ag_2F	Ag (0.5)
	$La_{1-x}Sr_xMnO_3$	Mn($3+x$)
	$K_2Pt(CN)_4Br_{0.3}3H_2O$	Pt (2.3)
	$Hg_{2.67}AsF_6$	Hg (0.37)

the basis of the useful classification scheme proposed by Day and Robin. Examples of compounds in each class are given in Table 6.1.

Class I This comprises compounds in which the different oxidation states are associated with very different environments. The energy required to transfer an electron between the two is large. Thus there is essentially no interaction between the different oxidation states, and no special properties associated with the mixed valency.

Class II These compounds also have different environments for the different oxidation states, but the sites are now sufficiently similar that electron transfer requires only a small energy. These compounds are semiconductors, and have optical absorptions resulting from the kind of *intervalence transition* described in the previous section. Examples of intervalence bands appearing in the optical spectra of Class II compounds of platinum and iron are shown in Fig. 6.16.

Class III These compounds are ones having all atoms in an identical, fractional oxidation state, with electrons delocalized between them. This category can be divided into two sub-classes: In Class IIIA the electron delocalization occurs within a finite cluster; Class IIIB is where the electrons

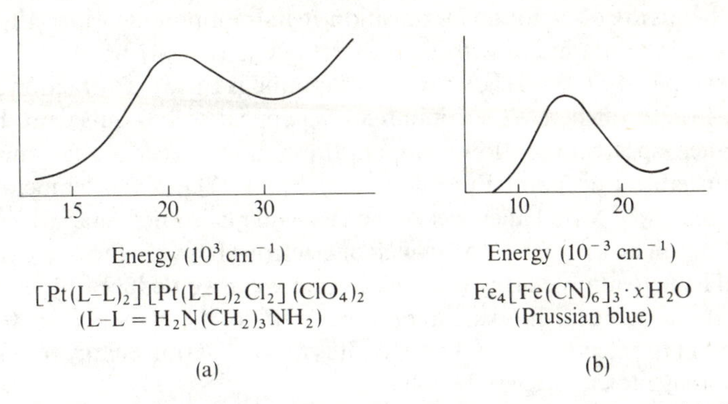

Fig. 6.16 Electronic spectra of some Class II mixed valency compounds, showing the intervalence band. (After R. J. H. Clark, in D. B. Brown (ed.), *Mixed-valence compounds*, D. Reidel, 1980); M. B. Robin, *Inorg. Chem.*, **1** (1962), 337.)

are delocalized throughout the solid. Compounds in this sub-class have metallic conductivity.

From our point of view the interesting question is: why does a given compound fall in Class II rather than in Class III? That is: what causes the *valence trapping* associated with Class II behaviour? Sometimes there are obvious structural features that make the different oxidation states more stable at particular sites. For example in Prussian blue, $Fe_4[Fe(CN)_6]_3 \cdot xH_2O$, the Fe^{2+} coordinates to the carbon end of CN^- and the Fe^{3+} to the nitrogen end. In the sodium tungsten bronze Na_xWO_3 with small x, the extra electrons introduced by the sodium are probably trapped near Na^+ sites in the lattice, where there is a favourable electrostatic potential; these compounds show a transition to Class III delocalized behaviour as x increases, which is analogous to that which takes place in concentrated alkali metal–ammonia solutions. Alternatively, the tungsten bronzes could be described as doped semiconductors, which will be discussed in Chapter 7. In many Class II compounds however, the lattice sites occupied by the different oxidation states are distinguishable *only* as a result of the valence trapping. The localization must then be caused by lattice distortion effects of the small polaron type.

6.3.2 Oxidation states differing by one

When the trapped oxidation states differ by one unit, we can consider that there are extra electrons or holes in the solid, with a small polaron trapping of the type discussed in Section 6.2. The main driving force for the trapping is the difference in ionic radii between the two oxidation states, which causes local lattice distortions. As we have seen however, the trapping energies are not large, and localization is only expected when the bands are rather narrow. Such narrow bands are often found in transition-metal compounds, especially the 3*d* series, and in lanthanides, with their extremely narrow 4*f* band.

An example of a Class II lanthanide compound is Eu_3S_4. Diffraction studies show a structure where all europium atoms appear to be equivalent, but the Mössbauer spectrum at low temperatures shows clearly that there are distinguishable Eu^{2+} and Eu^{3+} ions (see Fig. 6.17). As the temperature is raised, the two Mössbauer peaks coalesce. This is not due to electron delocalization however, but to *hopping* of electrons between the two oxidation states. The activation energy for such hopping can be deduced by fitting the spectra to theoretical models, and comes out to the same value (0.24 eV) observed in the electronic conductivity. It is this activation energy for electron transfer that indicates Class II behaviour.

The 3*d* bands in transition-metal compounds are less narrow than those of 4*f* orbitals, and this often leads to behaviour that is close to the borderline

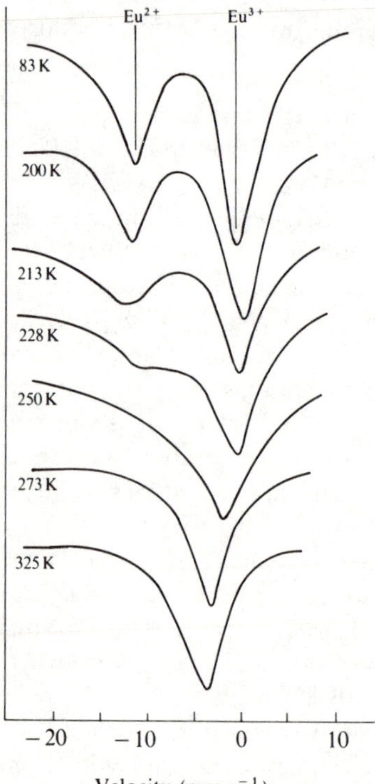

Fig. 6.17 Mössbauer spectrum of Eu_3S_4 at different temperatures, showing distinct Eu^{2+} and Eu^{3+} peaks, and their coalescence due to electron hopping. (After O. Berkooz *et al.*, *Solid State Commun.*, **6** (1968), 185.)

between Classes II and III. A well-known example is Fe_3O_4, in which both Fe^{2+} and Fe^{3+} are present on octahedral sites in the inverse spinel structure, $Fe^{3+}[Fe^{2+}Fe^{3+}]O_4$. The variation of conductivity with temperature is shown in Fig. 6.18. At room temperature, the conductivity is in the metallic range, although it does not behave quite as a metal should, showing a slow increase with temperature. It appears that the electrons are not quite free, and that their mobility is still controlled by lattice distortion effects. At 120 K, the so-called **Verwey transition** occurs, and the conductivity drops, showing an appreciable activation energy at lower temperatures. Although the structural details are very complicated, it is certain that the transition is associated with a cooperative distortion of the lattice, which traps the valencies by the formation of distinguishable Fe^{2+}–O and Fe^{3+}–O bond distances.

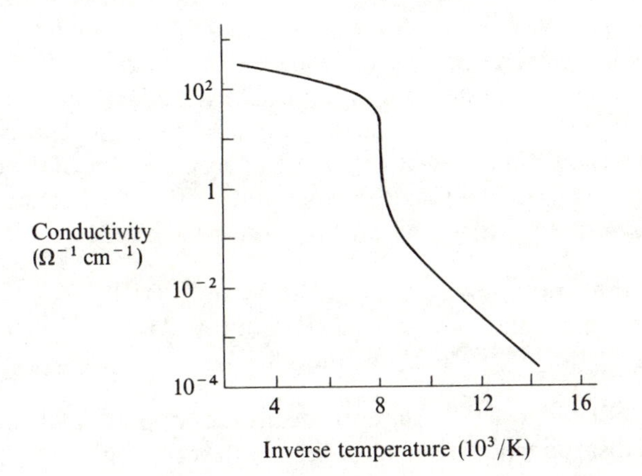

Fig. 6.18 Conductivity of Fe_3O_4 as a function of temperature. (After J. B. Goodenough, in D. B. Brown (ed.), *Mixed-valence compounds*, D. Reidel, 1980).)

Mn_3O_4 and Co_3O_4 show quite different behaviour from that of Fe_3O_4. The manganese and cobalt compounds crystallize in the normal spinel structure, where all the M^{3+} ions are in octahedral sites, and the M^{2+} is tetrahedral. Electron transfer between the different oxidation states in different lattice sites is energetically more difficult than in Fe_3O_4, so that the conductivity is much lower.

The way in which the electronic properties are influenced by mixed valency is illustrated clearly in the lanthanum strontium manganate system, $La_{1-x}Sr_xMnO_3$. The end members are both non-metallic, with electrons localized by electron repulsion. $LaMnO_3$ contains the Mn^{3+} ion with the high-spin $(3d)^4$ configuration, and shows a cooperative Jahn–Teller distortion rather similar to that described for Rb_2CrCl_4 in Section 5.3.3. When lanthanum is replaced by strontium, the different charge is compensated by oxidizing manganese to Mn^{4+}. With more than about 10 per cent Mn^{4+}, the compound becomes metallic, showing Class III mixed valence behaviour. At the same time, the Jahn–Teller distortion disappears, as a result of the delocalization of the e_g electrons. Another interesting change is that while the lanthanum and strontium compounds themselves are antiferromagnetic, the metallic mixed valency phase is ferromagnetic. This is due to the double exchange mechanism described in Section 5.3.4: electron transfer between Mn^{3+} and Mn^{4+} is more favourable if the spins on the neighbouring ions have a parallel alignment (see Fig. 5.18(c) on p. 158).

6.3.3 Oxidation states differing by two

Many compounds in this category appear at first sight to have atoms with integral oxidation states, as for example in $CsAuCl_3$ and Cs_2SbCl_6. The structural and electronic properties of these solids make it clear, however, that they are Class II mixed valency compounds, with Au(I, III) and Sb(III, V). The oxidation states found in this type of compound generally correspond to the stable states known in the solution and complex chemistry of the corresponding elements. Thus mixed valency compounds with oxidation states differing by two very often contain ions with the following electron configurations:

d^6 and d^8, both low spin, as in Pt(IV) and Pt(II)
d^8 (low spin) and d^{10}, as in Au(III) and Au(I)
$d^{10}s^0$ and $d^{10}s^2$, as in Sb(V) and Sb(III).

The intermediate oxidation state in these cases appears to less stable, and undergoes disproportionation according to the reaction:

$$2M^{n+} \rightarrow M^{(n+1)+} + M^{(n-1)+}. \qquad (6.11)$$

In the light of the discussion of electron repulsion effects in Chapter 5, such disproportionation is quite surprizing. In gas-phase ions the process (equation 6.11) is very unfavourable, and generally requires an energy input of at least 10 eV. In a solid, the energy required for disproportionation is the Hubbard U, responsible for localizing electrons in many transition-metal and lanthanide compounds. In the mixed valency compounds considered here, the process must be energetically *favourable*, and this is sometimes described as **negative *U***. The general occurrence of a negative U with certain electron configurations is not fully explained, but some of the factors involved are:

(i) The two oxidation states may have large differences of bond length or coordination geometry, so that the relaxation energies which localize the electrons are particularly large.
(ii) The atomic orbitals concerned are relatively diffuse, so that even in the gas phase, U is not too large. This may explain, for example, why the

disproportionation is often observed for elements in the second and third transition series, but not in the first.

(iii) The compounds all have a substantial degree of covalent character, which spreads out the electrons from their atomic orbitals, and so reduces the repulsion.

It is interesting to consider some examples from each of the electron configurations mentioned above.

Case 1: d^6–d^8. The best known examples here are platinum chain compounds with halogen bridges. Figure 6.19 shows the structure of Wolfram's Red Salt, $[Pt^{II}(et)_4][Pt^{IV}(et)_4Cl_2]Cl_4$ (et = aminoethane, $CH_3CH_2{\cdot}NH_2$). The chlorine atoms bridge the two types of platinum with distances differing by 30 pm, so that there is essentially an alternation of octahedral Pt(IV), and square planar Pt(II). These are the coordinations favoured by ions with the low-spin electron configurations d^6 and d^8 respectively. It is interesting that the hypothetical symmetrical case, with equal distances to Pt(III), would be a one-dimensional metal with a half-filled band. Thus the alternation of bond distances that accompanies the disproportionation could be regarded as a Peierls distortion of the kind discussed earlier in this chapter.

Mixed valence platinum chain compounds such as Wolfram's Red Salt owe their colour to intervalence bands, and the absorption spectrum of a similar compound was shown in Fig. 6.16. All the intervalence bands are rather broad,

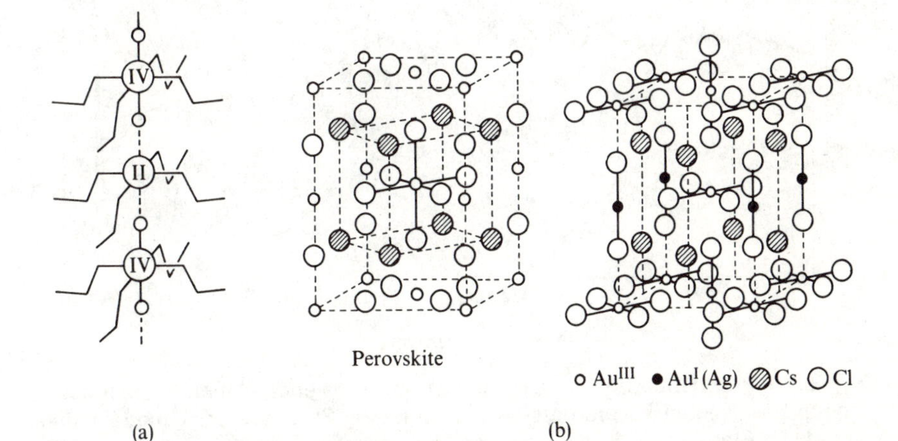

Fig. 6.19(a) Pt chain structure of Wolfram's Red Salt $[(C_2H_5NH_2)_4Pt][C_2H_5NH_2)_4PtCl_2]Cl_4$. (b) Structure of Wells' Salt, $Cs_2Au^IAu^{III}Cl_6$, showing its derivation from the perovskite lattice. (After R. J. H. Clark, in D. B. Brown (ed.), *Mixed-valence compounds*, D. Reidel, 1980); A. F. Wells, *Structural inorganic chemistry*, Oxford University Press, fifth edition, 1985).)

and this is due to the change in equilibrium geometry accompanying the electron transfer. According to the Franck–Condon principle, a change in geometry in the excited state is reflected in the vibrational modes excited in an electronic transition. We would expect the transfer of an electron from Pt(II) to Pt(IV) to shift the position of the bridging chlorine, and so to excite Pt–Cl stretching vibrations. Although the vibrational progressions cannot be resolved in the absorption spectrum, they can be seen in resonance Raman spectroscopy. Normally in Raman spectroscopy the optical exciting line is chosen to be at a wavelength where no absorption occurs. However, if a Raman spectrum is taken with light corresponding to an electronic absorption band, resonance effects occur, which can greatly enhance the vibrational bands. The modes enhanced are precisely those involved in the change in geometry of the electronic excited state. Figure 6.20 shows the resonance Raman spectrum of a Pt(II, IV) chain compound, excited in the intervalence band. The extensive vibrational progression can be identified as the Pt–Cl stretching frequency, and confirms that it is this bond length that changes most in the electronic transition.

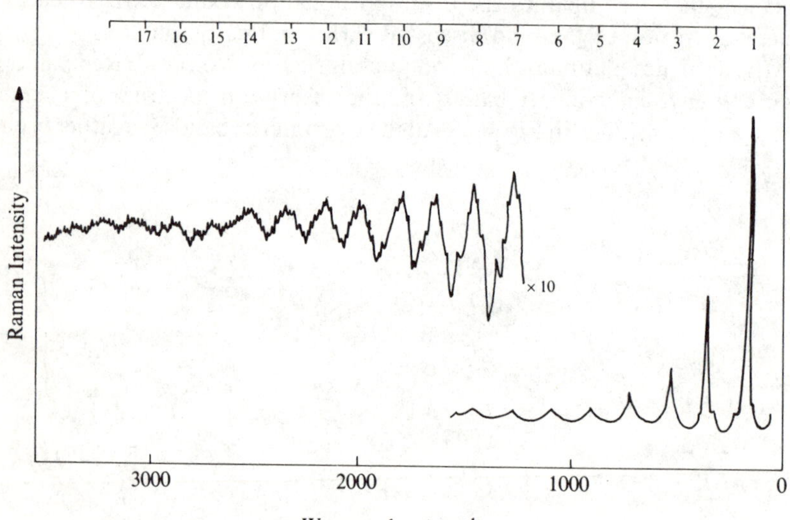

Fig. 6.20 Resonance Raman spectrum of mixed valence platinum chain compound, with extensive progression in the bridging Pt–Cl stretching mode. (After R. J. H. Clark, in D. B. Brown (ed.), *Mixed-valence compounds*, D. Reidel, 1980).

Case 2: d^8–d^{10}. Examples are $Ag^IAg^{III}O_2$, and Wells' Salt $Cs_2Au^IAu^{III}Cl_6$. Both compounds show the square planar coordination favoured by the d^8 ion, and a linear two-coordination commonly found with d^{10}. The structure of

Wells' Salt (Fig. 6.19(b)) shows how this coordination arises from a distortion of the perovskite structure, so that Au(III) is surrounded by four close chlorines and two further away, with the reverse for Au(I). Both this and Ag_2O_2 are black and semiconducting, which shows the rather low energy required for the intervalence electron transfer.

Case 3: s^0–s^2. These configurations are often found with post-transition metals, such as Pb(II, IV) and Sb(III, V). An example is Cs_2SbCl_6, which actually contains distinguishable $(Sb^{III}Cl_6)^{3-}$ and $(Sb^{V}Cl_6)^{-}$ ions, with Sb–Cl bond distances differing by 30 pm. The same ions are found when antimony is doped into the corresponding tin(IV) compound, Cs_2SnCl_6. Electron transfer can then occur between Sb(III) and Sb(V) if they are at neighbouring sites in the lattice. At low doping levels, the number of such neighbouring pairs should be proportional to the square of the antimony concentration. Both the electrical conductivity and the absorption strength of the intervalence band are found to obey this law. It is interesting that small concentrations of antimony doped into tin dioxide (SnO_2) show quite different behaviour. Instead of disproportionating, each antimony atom donates its extra valence electron to the conduction band of the solid, which becomes metallic.

Further reading

Various solids with one-dimensional electronic structure are reviewed in the following articles:

J. S. Miller and A. J. Epstein (1976). *Prog. Inorg. Chem.* **20** 1.
A. J. Heeger and A. G. MacDiarmid (1980). *Mol. Cryst. Liq. Cryst.* **74** 1.
J. M. Williams (1983). *Adv. Inorg. Chem. and Radiochem.* **26** 235
J. M. Williams (1985). *Prog. Inorg. Chem.* **33** 183.

Layer compounds with two-dimensional properties are described in:

A. D. Yoffe (1976). *Chem. Soc. Rev.* **5** 51.
J. A. Wilson, F. J. D. DiSalvo, and S. Mahajan (1975). *Adv. Phys.*, **24** 117.

The properties of solids where polarons are formed are discussed by:

D. Adler (1967). In *Solid state chemistry* (ed. N. B. Hannay), Vol. 2. Plenum Press.

The classic account of mixed valency compounds, setting out the widely used classification scheme described in the present chapter, is:

M. B. Robin and P. Day (1967). *Adv. Inorg. Chem. and Radiochem.* **10** 247.

More recent work in this field is reviewed in the following:

D. B. Brown (ed.) (1980). *Mixed-valence compounds*. D. Reidel.
P. Day (1981). *Int. Rev. Phys. Chem.* **1** 149.
C. Gleitzer and J. B. Goodenough (1985). *Structure and Bonding* **61** 1.

7
Defects, impurities, and surfaces

All solids, even the most 'perfect' crystals, contain defects and impurities, and have surfaces. These lead to a break in the regular periodicity of the crystal lattice, and to a perturbation in the electronic structure. The electronic and optical properties of many solids are in fact dominated by such effects. For example, the application of semiconductors in solid-state devices depends on the electronic levels of deliberately introduced impurities, that is on **doping**. The chemical interest in defects and surfaces extends far beyond such relatively simple solids however, and this chapter discusses their electronic consequences in a variety of different types of solid.

7.1 Structural and electronic classification of defects

7.1.1 Types of crystal defect

The early part of this chapter will for the most part be concerned with **point defects**. These are perturbations of the crystal that involve a single lattice site, or at most a small group of sites. The simplest examples are the **lattice vacancy** and the **interstitial atom** (see Fig. 7.1). Isolated impurities can also occupy either **substitutional** or interstitial positions. In ionic solids such elementary defects are often charged, since they introduce an imbalance of ions of different types. The charge on an individual defect plays an important role in its electronic properties. In the bulk of the crystal, a build-up of charge cannot be tolerated for more than a minute number of defects, as it would lead to a very high electrostatic potential. Significant numbers of charged defects can only exist, therefore, if the charge is *compensated* by that of other defects. The simplest combinations of defects in ionic crystals are shown in Fig. 7.1(c) and (d). The **Schottky defect** consists of an anion and a cation vacancy, with overall balancing of charge. In the **Frenkel defect**, on the other hand, a vacancy is balanced by an interstitial of the same type. Unlike Schottky defects, the Frenkel defect may be largely confined to one kind of ion; for example silver halides have a significant number of cation (Ag^+) Frenkel defects, but very few anion defects at room temperature. Similar considerations of charge apply to impurities. When one ion is substituted for another of different charge, it may

Fig. 7.1 Simple point defects. (a) Lattice vacancy. (b) Interstitial atom. (c) Schottky defect, consisting of cation and anion vacancy. (d) Frenkel defect, with vacancy balanced by interstitial.

be compensated by appropriate vacancies or interstitials in the host lattice. For example substitution of Ca^{2+} or Y^{3+} for Zr^{4+} in zirconia (ZrO_2) leads to the formation of compensating oxide vacancies.

Another way in which the charge on a defect can be compensated, and one that has more interesting electronic consequences, is by extra electrons or holes in the solid. This happens in semiconductors, when the extra nuclear charge of a substitutional phosphorus in silicon is compensated by an electron placed in the conduction band. A similar effect occurs when some tin is replaced by antimony in doped SnO_2. Another example is the *F-centre* in alkali halides, where an anion vacancy has a trapped electron that compensates the charge of the missing anion. The occurrence of extra electrons and holes is often associated with a change in oxidation state of one of the ions, and so this type of compensation is common in compounds of transition-metal elements which show variable oxidation state. In a transition-metal oxide for example, the extra electrons compensating the formation of oxygen vacancies are likely to enter the *d* band, and hence decrease the oxidation state of the metal. A metal vacancy on the other hand might be compensated by increasing the oxidation state of some of the remaining metal ions, introducing holes in the *d* band. The electronic properties of this type of defect will be discussed in more detail in later sections.

In addition to point defects, solids also contain **extended defects**, which can be classified as **linear** or **planar** according to the region of the crystal affected. The commonest type of linear defect is a **dislocation**, associated with a fault in the arrangement of atoms along a line in the crystal lattice. An example of a planar defect is the **shear plane** that occurs in some reduced oxides such as WO_{3-x}. Instead of forming point defects, the oxygen vacancies associated with the reduction congregate in a plane in the lattice, where the metal atoms are brought closer together. Another kind of 'planar' defect is a **grain boundary** between two crystallites in a polycrystalline ceramic. The most universal type of planar defect, however, is the crystal **surface**. An ideal surface forms a planar termination of the bulk crystal, although real surfaces may be very rough, and have their own point defects and impurities associated with them.

7.1.2 Electronic consequences of defects

All defects break the regular periodicity of the ideal crystal lattice, and as was seen in Section 4.1.6, this has the consequence that electron waves with different k values are mixed up, so that electrons travelling through the crystal are scattered into other orbitals. Thus defects and impurities in metals tend to decrease the electrical conductivity. The same is true in semiconductors, where electrons or holes thermally excited into the bands are also scattered by defects. However, in non-metallic solids defects and impurities can have much more important effects, since they can introduce extra electronic levels into the energy gap. The most important factors determining the electronic consequences of the defects are (a) the energies of the extra levels; and (b) the number of electrons that occupy them. Some of the different possibilities are illustrated in Fig. 7.2.

Figures 7.2(a) and (b) show cases where the defects have introduced extra electrons into the conduction band, or holes into the valence band. The solid then becomes metallic (for example, antimony-doped tin dioxide, or Na_xWO_3 with $x > 0.3$). More often, with small concentrations of defects, the electrons or holes are trapped in levels close to the band edges ((c) and (d) in Fig. 7.2). This is the situation in lightly doped semiconductors, such as silicon with parts per million of phosphorus or aluminium. Such defects not only give carriers that are easily freed and so increase the conductivity, but also change the Fermi level in the solid. The properties of doped semiconductors depend on both effects.

Figure 7.2(e) shows a case where a defect level just below the edge of the conduction band is empty in the ground state, so that it does not introduce extra electrons or alter the Fermi level. Levels such as this can act as traps for optically excited electrons. The trapped electrons may then form centres for

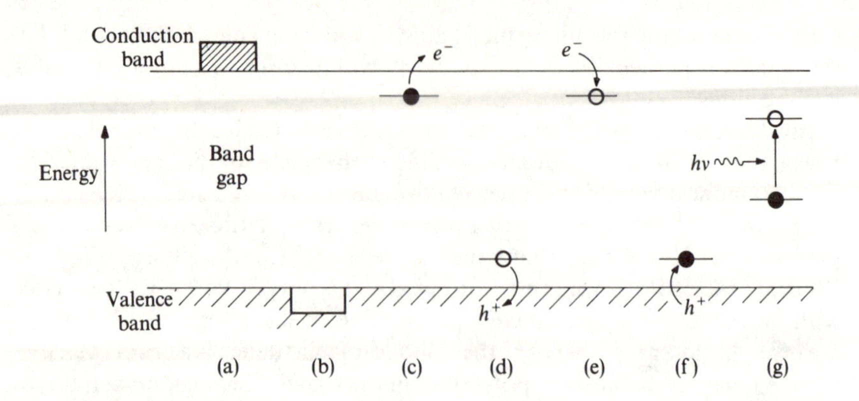

Fig. 7.2 Some electronic consequences of defects in non-metallic solids. (a) and (b): Extra electrons or holes in conduction or valence band. (c) and (d): Defect levels providing free electrons or holes by thermal excitation. (e) and (f): Defect levels acting as traps for electrons or holes. (g): Levels giving optical absorption at energies below the band gap.

interesting processes such as luminescence, or even chemical reactions such as those involved in photography (see Section 7.5.3). A similar defect is in Fig. 7.2(f), which is an occupied level just above the valence band edge, forming a trap for holes in the valence band.

Sometimes (Fig. 7.2(g)), the defect levels are rather far from the band edges, so that free carriers cannot be produced by thermal excitation at normal temperatures. This is the case with colour centres in alkali halides; as the name suggests, their most important property is that they have low-lying excited states, so that they can absorb light at energies well below the band gap of the perfect solid. Similar optical properties also result from impurities such as transition-metal ions, where the localized *d* electrons have their own optical absorption spectrum.

Most of the examples in the next few sections refer to point defects. Extended defects also have important electronic consequences. For example, dislocations in semiconductors increase the electrical resistance, and can act as centres for the recombination of electrons and holes. The shear planes in reduced oxides form traps for the extra electrons, so that the metal atoms with lower oxidation state are found there. However, the detailed electronic structure of extended defects is generally much less well understood than that of point defects. The electronic properties of surfaces, however, have been investigated intensively in recent years, and a few of the results will be discussed in the final section. Surfaces of non-metallic solids, especially when they are associated with defects or adsorbed species, show all the electronic phenomena summarized in Fig. 7.2.

7.2 Doped semiconductors

From a chemical point of view covalent semiconductors such as silicon and germanium are rather simple solids. The behaviour of impurities in these materials is well worth examining however, both because of its immense technological importance, and because the main principles involved can be extended to more complex solids, such as semiconducting oxides.

7.2.1 Donor and acceptor levels

When a Group V element such as phosphorus is doped into silicon, it enters the lattice as a substitutional defect, and replaces one of the tetrahedrally bonded host atoms. As we have seen, the valence band in silicon is filled by the bonding electrons. We would therefore expect the extra electron from each phosphorus atom to go into the conduction band. However, the tetrahedral phosphorus may be written P^+, with one greater nuclear charge than silicon. The electron in the conduction band will be attracted to this charge, and will 'orbit' round it in the ground state, rather like an electron in a hydrogen atom. This is called a **donor** state, with an electron trapped at an energy just below the conduction band. The situation is that illustrated in Fig. 7.2(c).

The energy of the donor state may be calculated roughly by the analogy with hydrogen, but there are three important modifications to be made to the normal equation:

(i) The energy zero corresponds, not to free space, but to an electron in the conduction band, at a large distance from the charge of the impurity. Thus the calculation will give the energy below the bottom of the conduction band (E_c).

(ii) The potential energy of attraction between the electron and the impurity will be reduced by the relative dielectric constant ε_r of the host. Thus

$$V(r) = -e^2/(4\pi\varepsilon_0\varepsilon_r r). \tag{7.1}$$

(iii) The kinetic energy of the electron will be modified by the fact that it is moving in the conduction band of silicon, and not in free space. As we have seen, the effect of the band structure can be taken into account by replacing the free electron mass m with an *effective mass* m^*. The modified free-electron formula for kinetic energy T in terms of the momentum p is:

$$T = p^2/2m^*. \tag{7.2}$$

The last two factors are very important in silicon, because it has a high dielectric constant ($\varepsilon_r = 12$) and electrons at the bottom of the conduction band have a low effective mass ($m^* = 0.2m$). The modified equation for the hydrogenic energy levels is:

$$E_n = -e^4 m^*/(32\pi^2 \varepsilon_0^2 \varepsilon_r^2 \hbar^2 n^2). \tag{7.3}$$

The most important energy is the ground state with $n = 1$, and equation 7.3 then gives the **donor ionization energy** E_d, that is the energy required to excite the electron to the conduction band. For silicon, the predicted value is 0.031 eV, and this may be compared with some experimental values (in eV) for donor levels of some Group V impurities: P, 0.045; As, 0.054; Sb, 0.047.

The agreement is not quantitatively perfect, one reason being that the hydrogenic model does not represent very well the actual potential field in the neighbourhood of the impurity atom. However, the calculation is qualitatively useful, and it is important to realize that one reason for its success is the large radius of the donor orbital. The formula for the Bohr radius of the ground-state 1*s* orbital is:

$$a_H = 4\pi\varepsilon_0\varepsilon_r\hbar^2/(e^2 m^*) \tag{7.4}$$

which for silicon is around 1 nm. The electron extends over many lattice spacings, and it is this that allows us, to a first approximation, to neglect the detailed atomic structure, and to treat the lattice as a continuous medium.

Silicon may also be substituted by a Group III atom such as aluminium. In this case, there is one less valence electron, and the result is an unfilled level, or hole in the valence band. The tetrahedral aluminium behaves as Al^-, with a nuclear charge of one less than silicon, and so the hole forms a similar bound state, an *acceptor level* just above the valence band. (If it is difficult to imagine a hole in this way, it may be helpful to think of the remaining electrons in the valence band as being *repelled* by the negative aluminium. Thus in the ground state, the unfilled orbital in the band will be close to the impurity.) A similar hydrogenic calculation gives the acceptor ionization energy, which is the energy required to ionize the hole into the valence band. (See Fig. 7.2(d): alternatively, it can be thought of as the energy required to excite an electron from a distant part of the valence band into the unfavourable region round the defect. This excitation leaves a hole free to conduct current in the valence band.) The predicted acceptor ionization energy for silicon is 0.044 eV, which is larger than that for donors because the holes in the valence band have a higher effective mass. Some experimental values (ev) are: B, 0.045; Al, 0.068; Ga, 0.071. Again the agreement is fair.

The rather low impurity ionization energies in silicon, which are very important in its electronic properties, can be seen from equation 7.3 to depend on:

(i) the high dielectric constant ε_r, which is a consequence of the small band gap (see Section 3.2.3 on p. 60); and
(ii) the low effective mass m^*, which is a consequence of the broad bands.

These features are unique to narrow-band-gap semiconductors such as silicon, germanium, and a number of compounds with low ionic character, such as GaAs and PbTe. With more ionic compounds, the dielectric constant is usually lower, and the bands are narrower so that the effective mass is higher. Impurity levels in these compounds are further away from the band edges, so that electrons and holes are not so easily ionized. At the same time, however, the success of the hydrogenic model for defect levels depends on the large radius of the orbitals, and this can be seen from equation 7.4 to depend on precisely the same factors which make the ionization energy low. Defects such as F-centres in ionic crystals have much more localized orbitals, and the hydrogenic formula is not so useful for calculating their energies.

7.2.2 Carrier concentrations and Fermi levels

The electronic properties of doped semiconductors are controlled by the concentrations of carriers (electrons and holes) and the position of the Fermi level. As was seen in Section 1.4.3, in a pure solid the Fermi level is approximately in the middle of the band gap:

$$E_F = (E_v + E_c)/2, \tag{7.5}$$

and the number of electrons (n) and holes (p) present at a given temperature is:

$$n = p \propto \exp(-E_g/2kT). \tag{7.6}$$

In these equations E_v and E_c are the energies of the edges of the valence and conduction band, respectively, and E_g the gap between them (see Fig. 7.3).

A solid with equal numbers of electrons and holes is known as an **intrinsic semiconductor**, since the properties are determined by the pure material. In doped semiconductors, one type of carrier usually predominates: thus we have **n-type** semiconductors, where extra electrons are provided from donor levels, or **p-type**, with extra holes from acceptor levels. The detailed calculation of the carrier concentration and Fermi level is quite complicated, but simple formulae can be given for certain limiting conditions. We shall assume for the moment that we have an n-type semiconductor with a low dopant concentration. This means that the concentration of donor electrons (n_d) is small compared with the density of states in the conduction band (N_c). The results are illustrated schematically in Fig. 7.3.

At *very low temperatures* most electrons are in the ground state donor orbitals, and very few in the conduction band. The same argument as for the intrinsic case (see Section 1.4.3 on p. 20) shows that the Fermi level should be mid-way between the donor energy (E_d) and the conduction band edge (E_c):

$$E_F = (E_d + E_c)/2. \tag{7.7}$$

The carrier concentration in the conduction band is given in an analogous way

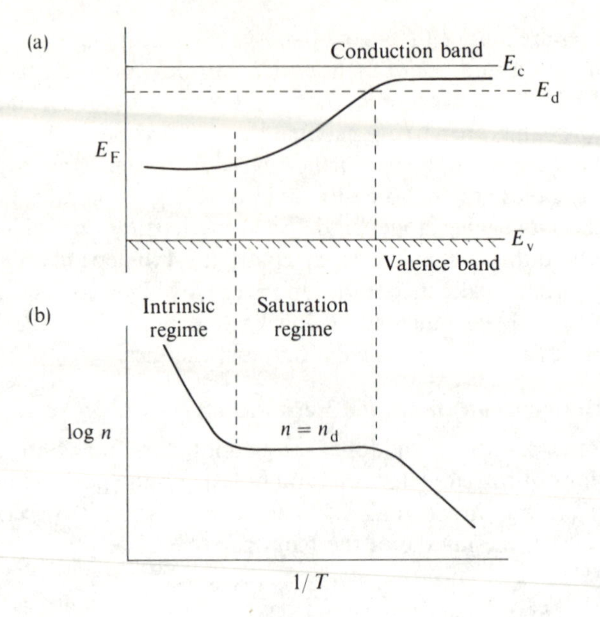

Fig. 7.3 Variation of (a) Fermi level, and (b) log of carrier concentration with inverse temperature in an n-type semiconductor.

by:

$$n \propto \exp[-(E_c - E_d)/2kT]. \tag{7.8}$$

As the temperature rises, n increases, and the donor levels themselves become depleted. A point is eventually reached when essentially all electrons are ionized into the conduction band:

$$n = n_d. \tag{7.9}$$

This is called the **saturation** or **exhaustion regime**, because an increase of temperature no longer gives an increase in carrier concentration. Silicon and germanium doped with Group V impurities are in the saturation regime at room temperature: although the donor ionization energy is still quite large compared with kT, the fractional occupancy n_d/N_c of the conduction band is small, because of the low concentration n_d of donor electrons. The Fermi level in the saturation regime can be calculated by equating this fractional occupancy to the Fermi–Dirac function:

$$n_d/N_c = 1/\{1 + \exp[(E_c - E_F)/kT]\}. \tag{7.10}$$

The exponential term is much greater than one, and so to a good approximation:

$$n_d/N_c = \exp[-(E_c - E_F)/kT]$$

whence:

$$E_F = E_c - kT \ln(N_c/n_d). \tag{7.11}$$

Equation 7.11 shows that in the saturation regime the Fermi level falls gradually with rising temperature, as the carrier concentration remains constant. However, a point is reached when the Fermi level comes close to the middle of the band gap. Now electrons start to be excited across the gap from the valence band, and eventually, at high enough temperatures, these electrons come to predominate over ones from the donor levels. Thus we reach the **intrinsic regime** when the doping is no longer important compared with carriers excited across the gap. The numbers of electrons and holes are nearly equal, and given by equation 7.6.

For a p-type semiconductor, the variation in hole concentration p follows the same pattern. In this case, the Fermi level at low temperature is mid-way between the valence band edge and the acceptor levels:

$$E_F = (E_v + E_a)/2 \tag{7.12}$$

and in the saturation regime is controlled by the number of acceptor levels n_a, and the density of states in the valence band N_v:

$$E_F = E_v + kT \ln(N_v/n_a) \tag{7.13}$$

Thus it rises with increasing temperature, and reaches the mig-gap position in the intrinsic regime.

Unless the band gap of the semiconductor is very small, or the impurity concentration very low indeed, the intrinsic regime is only reached at high temperatures. Under normal conditions, the Fermi level is in the region of the defect levels, that is close to the conduction band edge in an n-type semiconductor, or close to the valence band edge in the p-type case. This property of the defect levels is known as **pinning** the Fermi level, and it is found with many types of defect and surface levels in non-metallic solids. The position of the Fermi level in doped semiconductors is crucial to their applications, as will be seen in Section 7.2.4.

7.2.3 Electronic properties

Electrical conductivity is the most important property imparted by the electronic carriers. A general equation for conductivity may be written:

$$\sigma = n\mu_e e + p\mu_h e. \tag{7.14}$$

Here n and p are the carrier concentrations and μ_e and μ_h are their *mobilities*, defined in Section 4.1.6 (p. 101). The conductivity of a doped semiconductor shows approximately the same variation with temperature as the carrier concentration (Fig. 7.3). The difference is that the carrier mobilities decline slowly with temperature, due to scattering by thermally excited vibrations.

Thus in the saturation regime, where the carrier concentration is constant, the conductivity falls with increasing temperature.

A measurement of conductivity alone clearly does not allow the carrier concentration or type to be found. There are two important methods for giving an independent measurement of these. The first is the **Hall effect**, which was discussed in Chapter 4. The Hall coefficient is given by:

$$R_{\mathrm{H}} = -1/(ne) \tag{7.15}$$

in an n-type semiconductor, and:

$$R_{\mathrm{H}} = +1/(pe) \tag{7.16}$$

for a p-type material. When one carrier dominates, a measurement of the Hall effect thus gives a direct measure of its charge and concentration. Combining this with equation 7.14 then allows the mobility to be found. In fact, for a single carrier type (n $\gg$ p or vice-versa), we find:

$$\mu = |\sigma R_{\mathrm{H}}| \tag{7.17}$$

where μ is the appropriate mobility. Values determined in this way are called **Hall mobilities**. Because certain approximations are made in the theory, Hall mobilities may not agree perfectly with values measured by other methods.

The other important method for measuring the carrier type and concentration is the **Seebeck effect.** This is the phenomenon used in thermocouples: a potential is developed when two junctions between different materials are held at different temperatures, and is shown in Fig. 7.4 for a metal–semiconductor junction. The potential developed (ΔV) is related to the difference of Seebeck coefficients of the metal (S_{met}) and the semiconductor (S_{semi}):

$$\Delta V = \Delta T\,(S_{\mathrm{met}} - S_{\mathrm{semi}}). \tag{7.18}$$

Seebeck coefficients for semiconductors are usually much larger than those of metals. The main cause of the effect is the change of Fermi level with temperature. We shall look at the case of an n-type semiconductor in the saturation regime, where the Fermi level is given by equation 7.11. Figure 7.5 shows the situation. For simplicity, the Seebeck coefficient of the metal with

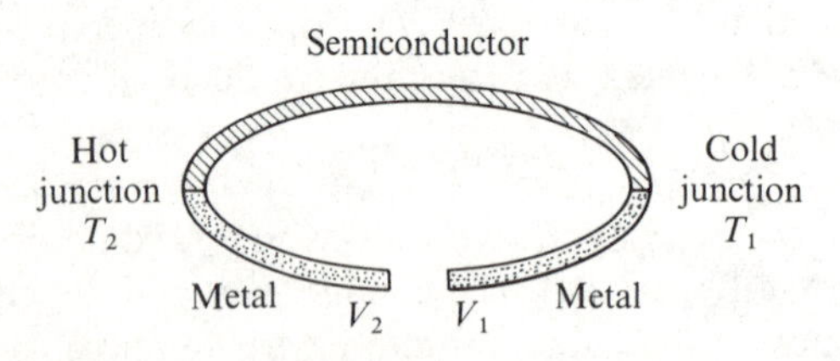

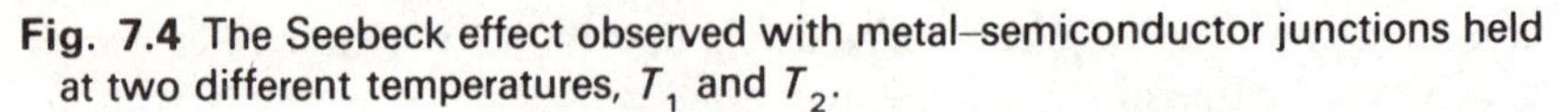

Fig. 7.4 The Seebeck effect observed with metal–semiconductor junctions held at two different temperatures, T_1 and T_2.

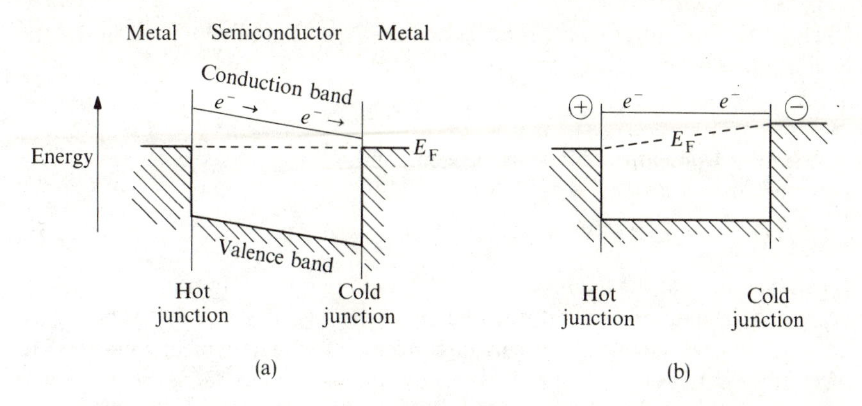

Fig. 7.5 Origin of the Seebeck effect in an n-type semiconductor. (a) No potential applied, with Fermi level constant throughout, and electrons flowing in semiconductor towards cold junction. (b) Voltage applied so that conduction band in semiconductor is kept level.

which the semiconductor forms the hot and cold junctions is assumed to be negligible. In Fig. 7.5(a) no potential has been applied between the two junctions, so that the Fermi level is constant throughout. However, from equation 7.11 the Fermi level on the hot side is lower relative to the band edges, and so the band energies must slope down towards the cold junction. Thus the carriers in the conduction band will flow 'downhill' between the hot and the cold junctions. (The junctions themselves are in equilibrium in Fig. 7.5, because in each case the Fermi levels of the metal and the semiconductor are the same. The Fermi–Dirac distribution in the metal extends to energies above the band edge of the semiconductor.) In order to equalize the band energies throughout the semiconductor, a negative potential must be applied to the cold junction, which raises the Fermi level: we can see from equation 7.11 that the energy change required is:

$$\Delta E = -k\Delta T \ln(N_c/n_d)$$

so that the potential which must be applied is:

$$\Delta V = -(k/e)\Delta T \ln(N_c/n_d). \tag{7.19}$$

There is another term in the Seebeck coefficient, however, since even with the bands level as in Fig. 7.5(b), the electrons are not in equilibrium. Electrons on the hot side of the semiconductor will have a larger average kinetic energy than those on the cold side, and so have more tendency to diffuse towards the other junction. To prevent this, an additional potential must be applied. The value is difficult to calculate, because it depends on the way in which the mobility of the electrons increases with their energy. However, this so-called **kinetic term** is independent of the electron concentration, and is of the same sign as the more important term (Equation 7.19). The final equation for the Seebeck coefficient can be written:

$$S = -(k/e)\,[\ln(N_c/n_d)+C]. \tag{7.20}$$

A similar calculation for a p-type semiconductor gives a similar result, but with the opposite sign:

$$S = +(k/e)\,[\ln(N_v/n_a)+C]. \tag{7.21}$$

It can be seen that the Seebeck effect provides an immediate test of the carrier type, and in a semiconductor where the densities of states N_c and N_v are known, a measurement of the carrier concentration. These measurements are often simpler to perform than those of the Hall effect, and the Seebeck effect is often used for routine testing of doped semiconductors. Typical values for S are in the range 0.1 to 1 mV/K.

7.2.4 p–n junctions

The fundamental component of solid-state devices that use doped semiconductors is the **p–n junction**, illustrated in Fig. 7.6. Such a junction may be made by diffusing a dopant of one type into a layer of a semiconductor of the other type. If the bands are level (Fig. 7.6(a)), the junction cannot be in equilibrium, because the Fermi levels on the two sides are not the same. Electrons will pass from the n- into the p-type material, forming a **space-charge region** where there are no carriers. The unbalanced charge of the ionized impurities causes the bands to bend, until a point is reached where the Fermi levels are equal (Fig. 7.6(b)).

The simplest property of the p–n junction is that of **rectification**, that is of passing current much more easily in one direction than the other. The depletion of carriers from the junction region forms an effectively insulating

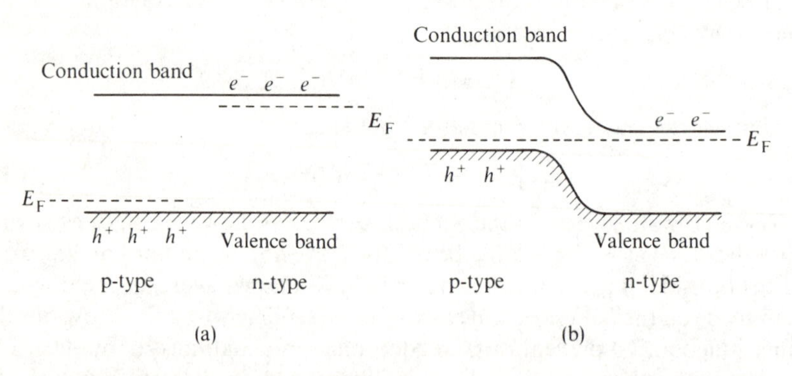

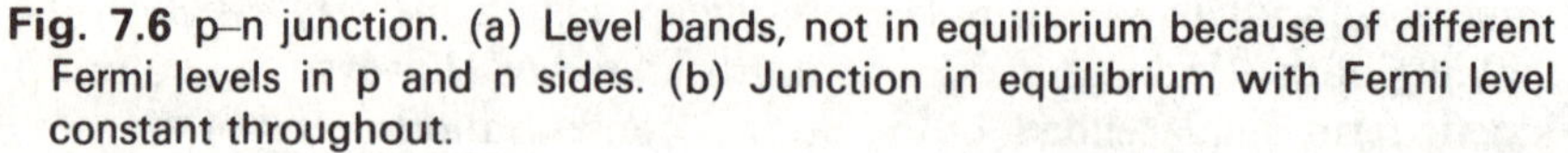

Fig. 7.6 p–n junction. (a) Level bands, not in equilibrium because of different Fermi levels in p and n sides. (b) Junction in equilibrium with Fermi level constant throughout.

barrier between the two sides. If a positive potential is applied to the n-type side—the situation known as reverse bias—more carriers are removed, and the barrier becomes wider. This is shown in Fig. 7.7. However, under forward bias, when the n-type side is made negative relative to the p-type side, the energy barrier for electrons and holes at the junction is decreased, and so carriers may flow through the junction, as shown.

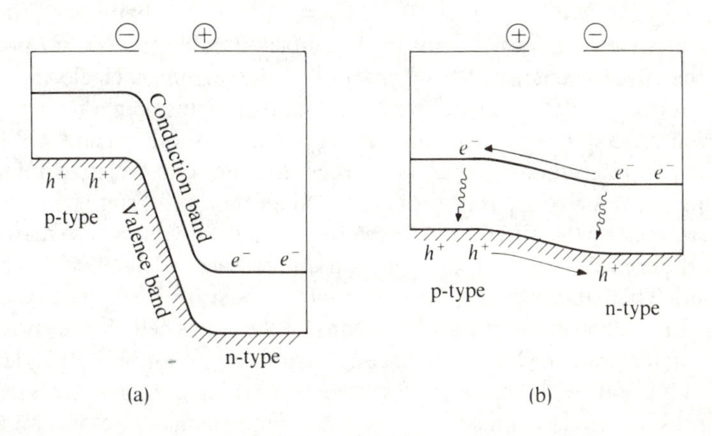

Fig. 7.7 Rectification by p–n junction. (a) Reverse bias applied: energy barrier and width of depletion region increased (b) Forward bias, showing passage of carriers, and their subsequent recombination.

Electrons passing into the p-type semiconductor recombine with holes, and the same happens with holes reaching the p-type side. Two mechanisms for recombination are:

1. **Non-radiative**: the energy of recombination passes into lattice vibrations, and so appears as heat.
2. **Radiative**: a photon is emitted, with an energy corresponding to the band gap of the semiconductor.

In normal semiconductor devices, made of doped silicon, the recombination is mostly non-radiative. The band gap in silicon is indirect (see Section 4.3.1), so that the radiative recombination is forbidden. However, in some materials such as gallium arsenide and phosphide (GaAs and GaP) which have direct gaps, a fair proportion of radiative recombination occurs. This kind of recombination can also occur at impurity sites. p–n junctions can then act as **light-emitting diodes** (LEDs), since they emit light when current is passed under forward-bias conditions. Some light-emitting semiconductor junctions can also act as lasers, where the radiative recombination of electrons and holes is stimulated by an incoming photon.

The reverse of light emission is the **photovoltaic effect** used in simple solar cells. Photons with energy greater than the band gap of a semiconductor can be absorbed to produce electrons and holes. Normally these recombine rapidly, but when light is absorbed in a p–n junction, the slope of the bands causes the electrons and holes to move in opposite directions, so that they separate (see Fig. 7.8). Electrons pass into the n-type side, and holes to the p-type, so that recombination is prevented. The carriers can be collected by metal electrodes and used to do work in an electrical circuit. When the band-bending in the junction corresponds to the normal equilibrium value there is no potential, since the effective energy of electrons and holes reaching each electrode is the same. In order to do electrical work, the photocell must generate a potential. This will reduce the band-bending in the junction. Some current will flow so long as the potential developed by the cell does not completely eliminate the bending necessary to separate the carriers. Since the equilibrium band-bending is determined by the difference of Fermi levels in the n- and p-type materials, it is this difference that controls the maximum voltage obtainable from a single junction. The band gap of the semiconductor must obviously be low enough for a reasonable proportion of photons to be absorbed. Since most solar radiation reaching the Earth's surface corresponds to photon energies less than 3 eV, only solids with band gaps less than this can be used in solar cells. Most solar cells are made from silicon, but compound semiconductors can also be used, for example cadmium selenide (CdSe) with a band gap of 1.8 eV.

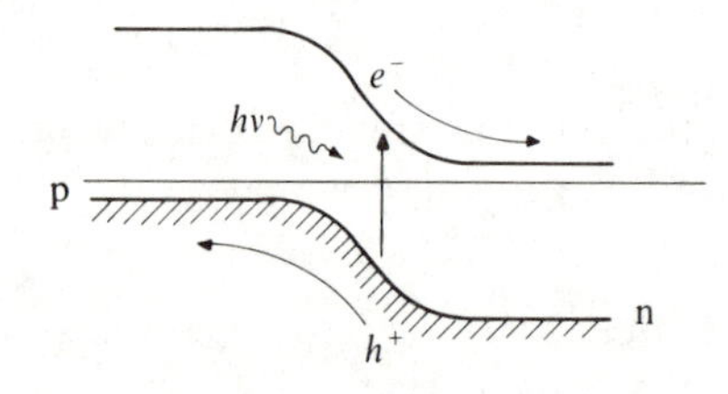

Fig. 7.8 Photovoltaic effect. Electrons and holes produced by absorption of light in the junction are separated by the band bending.

7.2.5 Transition to the metallic state

At the very low doping levels used in semiconductor devices, the impurities are on average well separated from each other. At higher concentrations, the donor or acceptor orbitals will start to overlap with one another, to form an **impurity band**. An impurity band just below the conduction band in a heavily doped n-type semiconductor is shown in Fig. 7.9. Since each donor atom provides one electron, the impurity band will formally be half full. This does not immediately lead to metallic conduction, however, since as explained in Chapter 5, electrons in narrow bands are localized by repulsion effects. The

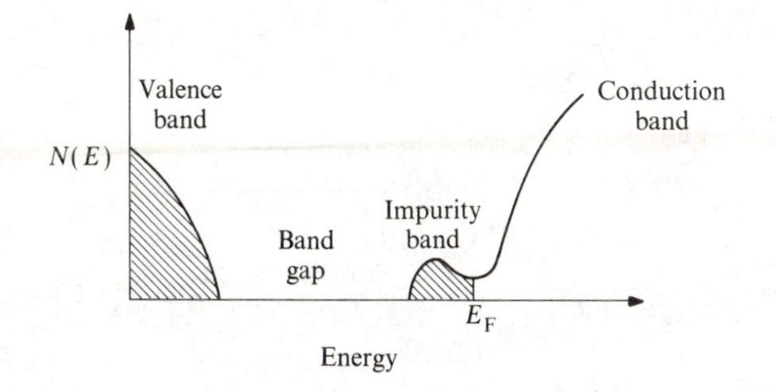

Fig. 7.9 Density of states in a heavily-doped n-type semiconductor, showing impurity band below the conduction band.

Hubbard criterion for metallic behaviour is that the width of the band (W) should be greater than the repulsion energy (U) for two electrons in the same orbital. The Hubbard U is related to the size of the orbital; the width of the band comes from the overlap of orbitals and depends on their separation. The mean separation of donor atoms in turn depends on their concentration, n_d:

$$\langle R \rangle \approx n_d^{-1/3} . \tag{7.22}$$

Mott showed that the solid should become metallic when $\langle R \rangle$ is proportional to the hydrogenic radius of the donor orbital, given by equation 7.4. The transition is expected when:

$$\langle R \rangle \approx 4a_H .$$

From equation 7.22 the **Mott criterion** is usually written:

$$n_d^{1/3} a_H \approx 0.25 . \tag{7.23}$$

A similar equation may be derived from the polarization catastrophe model described in Section 5.4.2. This predicts a transition when:

$$R/V = 1 .$$

The molar refractivity (R) of a donor electron is proportional to a_H^3, and the molar volume (V) to $1/n_d$. In this theory therefore, the transition also depends on the product on left-hand side of equation 7.23, although a detailed calculation leads to a slightly different numerical factor.

For phosphorus in silicon, the transition is found at a concentration of around 10^{19} atoms per cm^3, or one part in 10^4. Edwards and Sienko have found that the Mott criterion is satisfied quite well for systems undergoing metal–semiconductor transitions over a concentration range spanning a factor of nearly 10^9. Their remarkable plot is shown in Fig. 7.10. The higher concentration range includes systems such as expanded metals, and liquid ammonia solutions of alkali metals. The most important factor determining

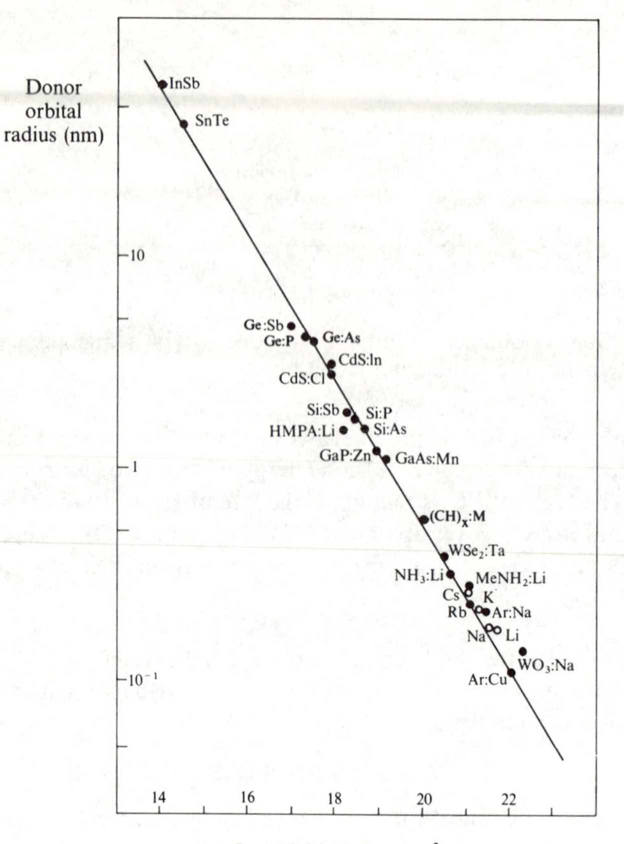

Fig. 7.10 Log–log plot of critical concentration required for metallic behaviour, against estimated radius of donor orbital. The line represents an empirical best fit to the data. (After P. P. Edwards and M. J. Sienko, *J. Am. Chem. Soc.*, **103** (1981), 2967.)

the radius of an impurity orbital is the dielectric constant of the medium. Semiconductors such as InSb and SnTe have very small band gaps and large dielectric constants. Thus a_H is very large, and they become metallic with very low impurity concentrations. On the other hand, rare gas matrices such as argon have low dielectric constants, and hence very contracted impurity orbitals. In these systems, therefore, much larger concentrations of dopants are required before their orbitals can overlap sufficiently for metallic conduction to occur. It is doubtful whether calculations based on hydrogenic orbitals should be used for such systems, but even so, the Mott criterion seems to provide a useful qualitative guide to their behaviour.

7.3 Defects in ionic solids

Ionic solids such as alkali halides have smaller dielectric constants than the covalent semiconductors discussed in the previous section. Electrons are therefore bound much more strongly to charged defects, and the hydrogenic approximation is no longer very useful. The ground-state electronic levels are generally quite well localized in the region of the lattice defect.

7.3.1 F-centres in alkali halides

An **F-centre** consists of an electron trapped at a halogen vacancy. These defects can be produced by heating the halide in an alkali metal vapour, and their name (from the German *Farbe*, which means colour) comes from the colour produced by the optical absorption spectrum of the trapped electron. The binding of the electron to the vacancy is a result of the net positive charge left by the vacant halide site. Figure 7.11 shows a rough picture of the ground-state electron distribution in the defect, and some of the approximate models that have been used to describe it.

The simplest theory of the F-centre is the **electron-in-a-box model**. This assumes that the electron is completely trapped in the vacant site. The optical absorption results from the excitation of the electron from its ground state, to excited state orbitals in the box. The exact formulae for the states of an electron in a spherical box are complicated, but like the cubic box (used in the free-electron theory in Chapter 3), the energies are proportional to $1/a^2$, where a is the size of the box. In fact, an empirical study of the absorption energies of F-centres in a variety of halides gives the approximate relation:

$$E = C\,a^{-1.7}\,. \qquad (7.24)$$

This is not too far from the electron-in-a-box prediction, but the deviation shows that the electrons do spread out somewhat from the defect region. The detailed electron distribution can be studied by ESR, or by a more sophisticated electron–nuclear double resonance technique (ENDOR). The spectra show hyperfine coupling due to the electron density on atoms surrounding the vacant site. With F-centres in KBr, for example, hyperfine splittings can be detected for atoms as far away as the sixth shell of ions from the vacancy, although some 63 per cent of the electron density is probably confined within the first shell of ions, and 99 per cent within the first three shells.

The fact that the electron distribution is not strictly confined to the lattice vacancy means that more realistic models are required for a proper treatment of the F-centre. Two of these are also shown in Fig. 7.11. They are:

1. Solution of Schrödinger's equation for the electron in a more realistic potential well, which can be calculated as a function of distance from the vacancy. This theory is quantitatively quite successful in calculating the electron distribution and excitation energies.

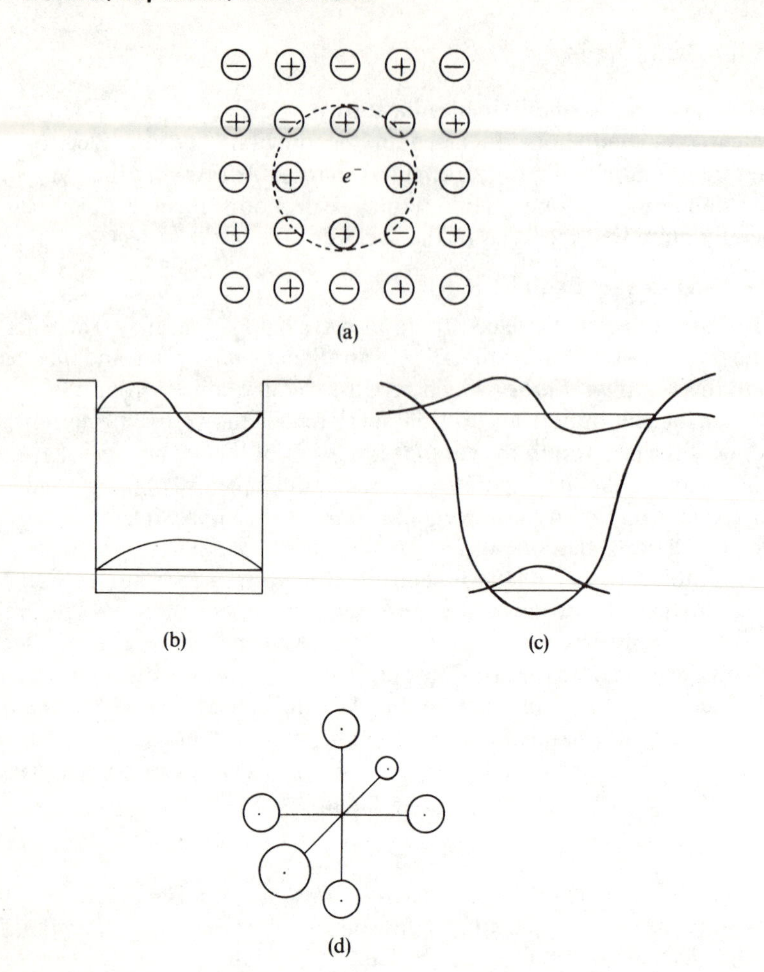

Fig. 7.11 F-centre in an alkali halide MX, and some theoretical models. (a) Approximate distribution of ground-state electron density in halogen vacancy; (b) electron-in-a-box model; (c) more realistic potential well; (d) LCAO wave functions constructed from *s* orbitals of cations surrounding the vacancy. Ground and excited state energies and wave functions are shown in (b) and (c).

2. An LCAO approach, writing the defect wave function as a linear combination of vacant atomic orbitals on ions surrounding the defect.

Figure 7.12 shows the F-centre absorption band in KBr at different temperatures, together with the spectra found for emission of light, when the F-absorption band is irradiated. Shifts of frequency between absorption and fluorescence bands are familiar in molecular spectroscopy, but that occurring

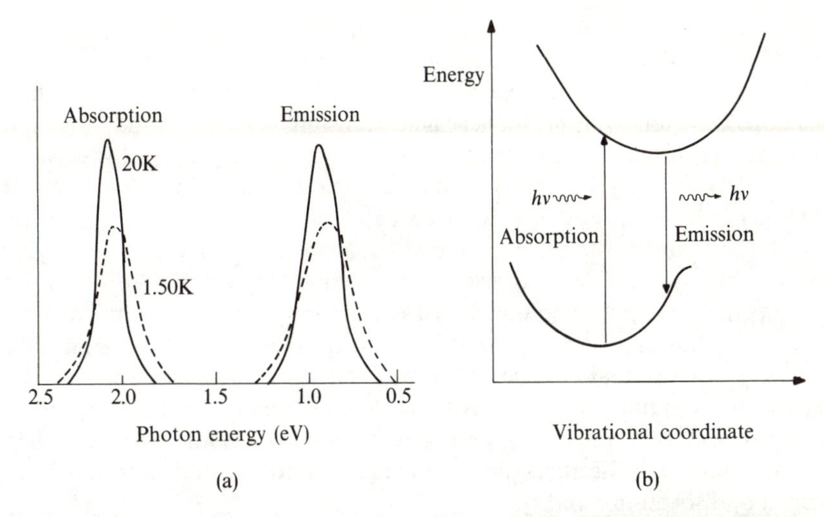

Fig. 7.12(a) Absorption and emission from F-centres in KBr at different temperatures. (After Gebhardt and Kuhnert, *Phys. Lett.*, **11** (1964), 15.) (b) Potential curves for ground and excited states, showing the origin of the Stokes shift between absorption and emission spectra.

with F-centres is abnormally large, and must come from a large change in equilibrium geometry of surrounding ions when the electron is excited. Figure 7.12(b) shows this. There is a large shift in the minimum position of the excited state potential curve, when plotted against the atomic positions. By the Franck–Condon principle, the electronic excitation is accompanied by a strong vibrational excitation. The vibrational energy of the excited state is quickly lost to the lattice however, and so emission occurs mostly from the ground vibrational level of the excited state curve, and is at considerably lower energy than absorption. The vibrational excitation also explains the width of the bands in the spectrum, and the way in which this changes with temperature. Because of the range of vibrational frequencies excited, individual lines are not resolved, but merely give a broadening of the spectrum. The width of the bands increases with temperature, because of the thermal population of vibrations in the ground state. The theory of vibrational broadening shows that the width should be given by:

$$W \approx h\nu[S \coth(h\nu/2kT)]^{1/2} \tag{7.25}$$

where ν is the vibrational frequency, and S the average number of vibrational quanta excited by the change in geometry in the electronic transition. For F-centre absorption bands, S is typically around 20.

The large change in equilibrium geometry indicates that the distribution of the electron itself must be different in the excited state. The more realistic potential well model described above (see Fig. 7.11(c)), suggests that the excited state orbital will be much more diffuse than the ground state. As a result, the ions feel the potential of the vacancy more strongly, and relax their positions accordingly. Ionic model calculations predict a very large distortion of the lattice around a 'bare' vacancy, which can give an energy lowering of several electron volts. This relaxation is important in the formation of defects in ionic crystals, since otherwise the energies required would be prohibitive.

At higher temperatures, irradiation of the F-absorption band does not give rise to fluorescence, but instead to photoconductivity. This observation suggests that the electron in the excited state no longer remains trapped at the defect site, but can escape into the conduction band of the crystal. The relatively low thermal energies required for such a process show that the excited state is quite close in energy to the conduction band edge. As would be expected from this conclusion, photoconduction can also be observed when crystals containing F-centres are irradiated at photon energies slightly above the main absorption band.

The F-centre is only one of many defects observed in alkali halides. More complex centres may involve cation vacancies, or aggregates of several vacancies. Thus F_2, F_3 and further aggregates have been studied, where two, three, or more adjacent halogen vacancies have trapped electrons. Another interesting defect is the X_2^- centre, where X is a halogen. From the discussion in Section 6.2.1, the very small width of the valence bands in alkali halides suggests that holes in the band should be trapped by lattice distortions, to form small polarons. In fact, the trapping is slightly more complicated than the simple polaron model would suggest. A hole on a chloride ion makes a chlorine atom, which is very reactive, and combines with a neighbouring chloride ion to form a Cl_2^- species. These centres have also been studied by ESR and electronic absorption spectroscopy.

7.3.2 Semiconducting oxides

Simple oxides such as MgO can have defects similar to those found in alkali halides. Because of the higher lattice energies, however, the creation of such defects in oxides requires much more energy than in halides, and they are normally observed only after rather extreme treatment, such as bombardment with ionizing radiation. Many oxides of transition and post-transition elements, on the other hand, can have quite high concentrations of defects under equilibrium conditions. The defects are often associated with deviations from perfect stoichiometry, and are formed because of the relative ease of reduction or oxidation of the metal ion. These oxides have considerably higher dielectric constants than the alkali halides, and so the electrons or holes are not so firmly bound to the defects. Thus carriers can be thermally excited into the conduction or valence bands, giving semiconducting properties. Some examples of semiconducting oxides are shown in Table 7.1. They include both 'pure'

Table 7.1
Some semiconducting oxides

Compound	*Carrier*	*Principal defect type*
TiO_2	n	O vacancy or interstitial Ti
ZnO	n	Zn interstitial
SnO_2	n	O vacancy
NiO	p	Ni vacancy
Li-doped MnO	p	Li^+ substituting Mn^{2+}
Na_xWO_3	n	Na interstitial
$Li_xV_2O_5$	n	Li interstitial
H_xMoO_3	n	OH^- substituting O^{2-}

(although in fact slightly non-stoichiometric), and doped compounds. The activation energies for conduction in these solids are typically in the range 0.1–0.5 eV, showing that the carrier ionization energies are higher than those found in doped silicon (0.03–0.05 eV), but lower than the binding energies of defect electrons in alkali halides (1–2 eV). As we shall see, the activation energy for conduction in a metal oxide may also have a contribution from the mobility of the carriers themselves.

n-type behaviour in an oxide is a result of slight reduction, which leaves extra electrons in the solid. This is found with compounds such as TiO_2 (which can be reduced by heating in hydrogen), and ZnO and SnO_2 (which are naturally slightly deficient in oxygen at high temperatures). At higher temperatures, the electrons will be mobile in the conduction band, but in the ground state they are normally trapped by the lattice defect. Reduction can be accompanied by formation of interstitial metal atoms, or by oxygen vacancies. An interstitial cation in the lattice will give a positive potential that can trap an electron in a bound orbital just below the conduction band edge, similar to the impurity levels introduced by phosphorus in silicon. The unbalanced positive charge at an oxide vacancy likewise forms a trap just as in an F-centre.

In an analogous way, a slight oxidation of a compound will remove electrons, and hence introduce holes, which may be mobile in the valence band. The case of NiO illustrates this. The eight *d* electrons of Ni^{2+} are localized by electron repulsion effects, so that very pure nickel oxide is an insulator, with a band gap of 3.8 eV. However, it is oxidized slightly on heating in air, and forms some nickel vacancies. For each Ni^{2+} vacancy, the charge must be compensated by oxidation of two more Ni^{2+} ions to Ni^{3+}. As in a mixed valency compound, conduction can occur by transfer of an electron between Ni^{2+} and Ni^{3+}. This could be described as the motion of holes in the nickel *d* band, although as we have seen in Chapter 5, simple band theory is not really applicable here. In the ground state, it is electrostatically favourable for the Ni^{3+} to be adjacent to the unbalanced charge of the Ni^{2+} vacancy, so that the holes are effectively trapped in acceptor orbitals, like those formed with Group III impurity orbitals in silicon.

The carrier type (n or p) in a semiconducting oxide can be investigated by a number of different techniques. The Hall and Seebeck effects described in Section 7.2.3 have been applied to oxides. The Seebeck effect is often the more useful, as it is difficult to interpret Hall measurements on magnetic oxides with localized electrons. For example, the Hall coefficient of NiO apparently changes sign near the Néel temperature. The reason for this is not understood, and the Seebeck coefficient remains positive at all temperatures, as expected for a p-type material. There are two other ways of finding the carrier type. The first is by photoelectron spectroscopy. Binding energies of electrons in solids are usually measured from the Fermi level, and the position of the bands with respect to this level depends on the carrier type. Figure 7.13 shows photoelectron spectra of two ternary oxides, strontium titanate ($SrTiO_3$) and lanthanum vanadate ($LaVO_3$). The former is a d^0 compound, with the oxygen $2p$ valence band forming the top filled levels (see Section 3.4). It appears from the spectrum that the top of this band (corresponding to the minimum binding energy) is 3 eV below the Fermi level. This is just about the observed band gap in $SrTiO_3$, and shows that the Fermi level is close to the conduction band edge. From the discussion of Fermi levels earlier (Section 7.2.2), it can be seen that

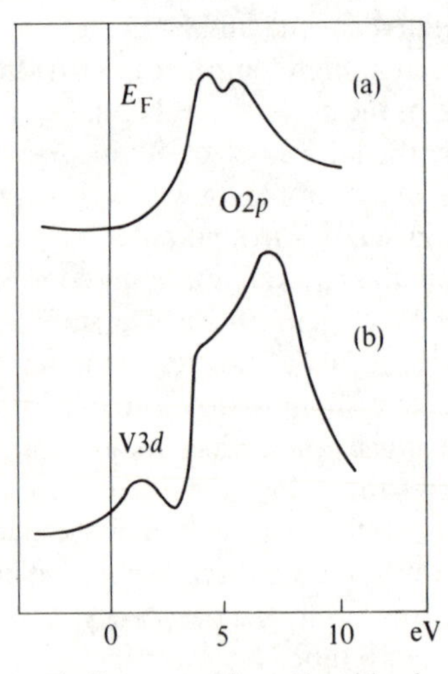

Fig. 7.13 Photoelectron spectra of (a) n-type $SrTiO_3$, and (b) p-type $LaVO_3$, showing the position of the top-filled levels with respect to the Fermi level.

this compound must be an n-type semiconductor. In fact, $SrTiO_3$ is easily reduced, and extra electrons are introduced, probably in conjunction with oxygen vacancies. The concentration of such donor electrons is too small to see in the spectrum; it is their influence in pinning the Fermi level that is important. $LaVO_3$ is a V^{3+} $(3d)^2$ compound, in which the $3d$ bandwidth is small, so that electrons are localized as in NiO. The V $3d$ electrons can be seen at lower binding energy from the oxygen $2p$ band. The Fermi level is now seen to be just at the onset of the spectrum, showing that it is at the bottom of the band gap, close to the occupied V $3d$ levels. Thus $LaVO_3$ must be a p-type semiconductor. This comes from some oxidation of V^{3+} to V^{4+}, compensating the presence of cation vacancies.

The other way of investigating a semiconducting oxide is to study the influence of oxygen partial pressure on the carrier concentration. In the n-type case, carriers arise from reduction of the oxide, which should be favoured by equilibrating the solid with low oxygen partial pressures. Thus the conductivities of the n-type oxides mentioned above are all found to fall with increasing oxygen pressure, as the temperature is kept constant. On the other hand, in p-type oxides such as NiO, the conductivity depends on oxidation of the solid, and rises with oxygen pressure.

Many transition-metal oxides can also be made semiconducting by doping, and some examples are included in Table 7.1. The general class of **oxide bronzes** can be included in this category. These are d^0 oxides such as V_2O_5 or WO_3, which are doped by elements such as alkali metals, or hydrogen. The alkali metals could be regarded as forming interstitials in the lattice, although this is not always a strictly accurate description, as the structure of the bronze may differ from that of the undoped oxide. Electrons from the dopants go into the conduction band of the oxide, and in some cases metallic solids are formed at high doping levels (for example in Na_xWO_3 for $x > 0.3$). However, at low concentrations, n-type semiconductors are usually formed. The electrons may be trapped partly by the potential of the alkali metal cation, but there is evidence in many of these compounds that the formation of small polarons is also a very important factor.

Substitutional doping is also possible, and will introduce carriers when the oxidation state of the dopant is different from that of the atom it replaces. For example, Li^+ can replace Mn^{2+} in MnO, and the change of charge must be compensated by the formation of some Mn^{3+}, which acts as a hole in the localized Mn^{2+} $3d$ levels, just as in NiO.

The simple chemical picture of semiconducting properties suggests that n-type behaviour should only be found in compounds where the metal can have an oxidation state lower than that in the pure solid. Conversely, p-type oxides are those where the metal can increase its oxidation state. Although this is usually the case, there are occasional surprizes. For example, some d^0 oxides such as $BaTiO_3$ show *p-type* semiconduction when they are prepared at high oxygen partial pressures. Neither of the cations Ba^{2+} or Ti^{4+} can be oxidized. It is believed that this behaviour is due to small concentrations of lower valence impurities. For example, Al^{3+} may replace Ti^{4+}, the charge being compensated by holes in the oxygen $2p$ valence band.

The detailed electronic behaviour of semiconducting oxides is often very complex, and is still an active field of research. In many cases, the precise nature of the defect sites is not understood. Another source of controversy concerns the nature of the 'free' carriers, when they are thermal ionized from the defect. It was seen in Chapter 6 that an electron or hole in an ionic solid inevitably produces some distortion of the surrounding lattice, and forms a polaron. When the polaron is large, the carrier can move in the conduction or valence band with no activation energy. In this case, the activation energy for conduction is simply controlled by the ionization energies of carriers from the defects. In other cases however, the carriers form small polarons, which hop through the lattice by an activated process. The activation energy for conduction is then the sum of the trapping energy at the defect, and the activation energy for the small polaron hopping. A separation of these two terms is not always easy. One approach is to use the Seebeck coefficient to measure the change in carrier concentration with temperature, and compare this with the variation in conductivity. It appears from this kind of study that holes in p-type MnO form small polarons with a hopping activation energy of about 0.4 eV. In NiO, however, the hole mobility does not appear to be activated, so that small polarons are not formed. Since the metal $3d$ bandwidths in these oxides are probably about the same, the difference must lie in the relaxation energies of ions round the M^{3+} site. This in turn can be traced (see equation 6.4 on p. 180) to the smaller high-frequency dielectric constant of MnO, which is more ionic than NiO.

7.4 Highly disordered solids

In the solids considered so far, the disorder is associated with defects that are relatively well isolated from one another. Some solids show such a high degree of lattice disorder, however, that this is no longer a valid picture. For example, many semiconductors can be made in an amorphous form, where there is no longer a crystal lattice with long-range order. The electronic structure of such amorphous materials has been extensively studied in recent years. Many of the gross features are quite similar to those of crystalline materials, but there are some subtle differences to be found in highly disordered solids.

7.4.1 Amorphous semiconductors

Semiconductors such as silicon and germanium can be made as amorphous films, for example by sputtering atoms onto a cold surface, or by the thermal decomposition of silane, SiH_4. Although the amorphous solids do not show long-range order like that in a crystal, there is considerable evidence, from diffraction and other studies, that the *local* coordination of most atoms is the same as in a crystal. In amorphous silicon, for example, most atoms are surrounded tetrahedrally by four others: the difference lies in the way in which these tetrahedra are linked together, giving an ordered array in the crystalline form, but not in the amorphous solid.

In Chapter 3 the principal features of the electronic structure of silicon—the valence and conduction bands and the energy gap—were shown to be a consequence of the tetrahedral bonding. The discussion did not depend on long-range order, and so we would expect the same features to be present in amorphous forms of silicon. This is confirmed, for example, by electronic absorption measurements and by photoelectron spectroscopy. Figure 7.14 compares the PE spectra of crystalline and amorphous silicon. The same features are shown in both spectra, although the disorder does seem to give some additional broadening of the bands in the amorphous form. Amorphous silicon is still a semiconductor, although various measurements show that there is a small concentration of occupied electronic states in the band gap. For example, the presence of unpaired electrons can be detected by ESR, and the conductivity is higher than that of pure crystalline silicon. These electrons are probably associated with silicon atoms that have not managed to achieve their preferred tetrahedral coordination during the growth of the amorphous film. A three-coordinate silicon atom would have a **dangling bond**. Such dangling-bond states are also known at crystal surfaces, and at lattice vacancies created

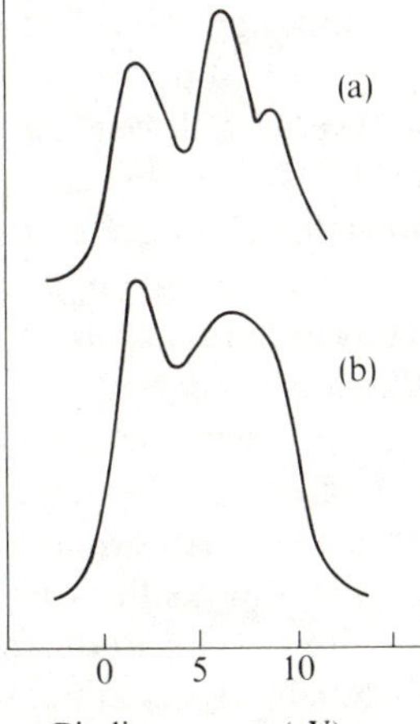

Fig. 7.14 Photoelectron spectra of (a) crystalline and (b) amorphous silicon. (After L. Ley *et al., Phys. Rev. Lett.,* **29** (1972), 1088.)

in crystalline silicon by high-energy irradiation. An electron in such a non-bonding orbital would be expected to have an energy intermediate between those of bonding orbitals, which form the valence band, and the antibonding orbitals which make up the conduction band. It is interesting that the presence of hydrogen, for example when amorphous silicon is made from SiH_4, greatly reduces the number of electrons in the band gap. It is likely that the dangling bonds are replaced by Si–H bonds, which have energy levels similar to those of the Si–Si bonds making up the valence band.

Another class of amorphous semiconductors is that of the **chalcogenide glasses**. These are compounds such as As_2S_3 and As_2Se_3, and in their crystalline modifications have structures where each arsenic is bonded to three chalcogens, and each sulphur or selenium to two arsenic atoms. Thus the filled valence levels are associated with bonding and non-bonding orbitals, and the conduction band with antibonding orbitals, as described in Chapter 3. Again, the amorphous solids probably have the same local coordination as the crystals, and so preserve the same basic electronic structure. It is interesting however that the chalcogenide glasses do not show dangling bond states like those of the amorphous Group IV semiconductors. There may be structural reasons why it is easier in the compounds for all atoms to achieve their ideal coordination in the glass, but it has also been suggested that dangling bonds might be stabilized by disproportionation: that is, the conversion of two singly-filled orbitals into a doubly-filled level and an empty one:

$$A\cdot + A\cdot \rightarrow A^- + A^+.$$

This disproportionation reaction is similar to one occurring in many mixed valency compounds, and involves the same **negative Hubbard *U*** (see Section 6.3.3). As in mixed valency compounds, such an electron transfer must be stabilized by a relaxation in the positions of surrounding atoms.

The disordered structure leads to lower electron mobilities, and the electrical properties of amorphous materials are inferior to those of single crystals. However, amorphous films do have some advantages, especially where a large area is required, so that a single crystal cannot be made, or would be prohibitively expensive. The alternative to an amorphous solid would then be a polycrystalline film, with many grain boundaries between the crystallites. Such grain boundaries can have a very deleterious effect on the electrical behaviour of a semiconductor, since they form traps for electrons and holes, and centres where they can recombine. Amorphous solids, on the other hand, can easily be made into large continuous films without grain boundaries.

One application for amorphous semiconductors is in **xerography**, the process used in many photocopiers. A metal-backed film of an amorphous semiconductor such as selenium or As_2Se_3 first has a positive charge deposited on the front. At room temperature, and in the dark, the semiconductor has a low conductivity, and the charge will remain for some time. The finely powdered ink adheres to the electrostatic charge, and thence can be transferred to the paper. When the semiconductor is exposed to light however, electrons and holes are created, giving rise to photoconductivity. The illuminated areas therefore lose their charge, and no ink adheres, thus giving the white regions in the copy.

7.4.2 Anderson localization

Although the gross features of electronic structure are not strongly affected by long-range disorder, there are more subtle effects in disordered solids, which can cause the electrons to be more localized than in a crystal. This phenomenon is known as **Anderson localization**. A rough idea of how it occurs can be seen in Fig. 7.15. An array of atoms is presumed to have a disordered potential field, so that the energies of the atomic orbitals vary randomly from

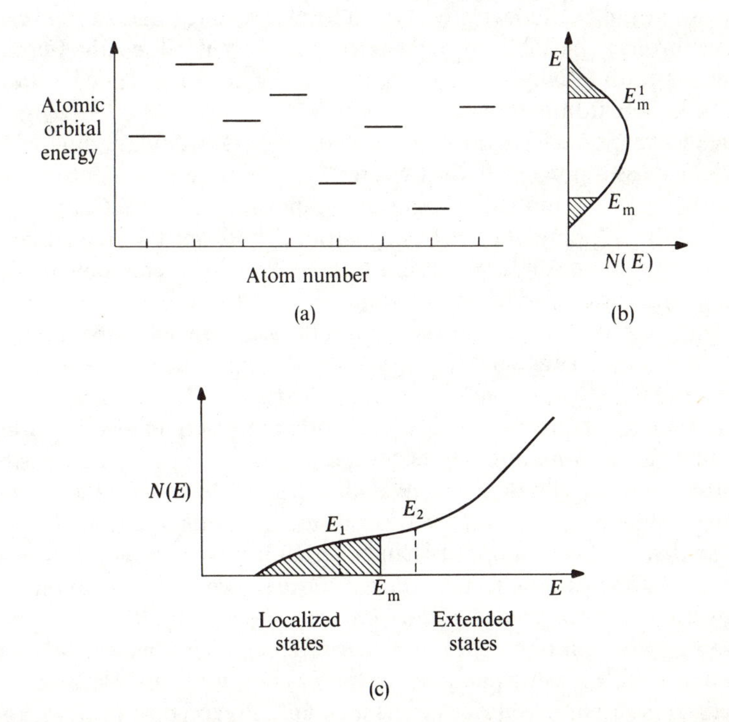

Fig. 7.15 Anderson localization in a disordered solid. (a) Array of atoms with randomly varying atomic energies. (b) Density of states, showing mobility edges E_m separating localized states at the edges of the band from extended states in the centre. (c) An Anderson transition occurs when the Fermi level moves from an energy (E_1) where states are localized, through a mobility edge to an energy (E_2) where extended states occur.

site to site. Atoms with near-average energies are very likely to have some neighbours at similar energy, and can form delocalized orbitals by their overlap. Atoms at exceptionally high or low energies however, are unlikely to have similar neighbours, and so their orbitals may remain as isolated, or localized, states. This picture suggests that the band of energy levels created from our disordered array may have orbitals of two types (see Fig. 7.15(b)): those in the middle of the band extend through the solid as in a crystal, and those close to the top and bottom of the band are localized in the vicinity of a particular atom. The region of localized states in the band will depend on the degree of disorder; in an extreme case all the states may become localized.

When there is a boundary between localized and extended states, as in Fig. 7.15, this is known as a **mobility edge**. The existence of a mobility edge can give rise to a metal–insulator transition, the so-called **Anderson transition**, as the band is filled up with electrons. This is shown in Fig. 7.15(c). When there are very few electrons present, the Fermi level may be below the mobility edge. Electrons at the Fermi level are then in localized states, and cannot conduct current through the solid. At higher electron concentrations, the Fermi level may cross the mobility edge, so that electrons are in extended states, and metallic conduction is possible.

The localization effects causing the Anderson transition are different from those considered previously. In Chapter 5, we showed how electrons can be localized by their mutual electrostatic repulsion, and in Chapter 6, we discussed the trapping of electrons by lattice distortions that they themselves produce. The Anderson localization is caused by a disordered potential which is due to the basic structure of the solid, and not to the electrons themselves. Unfortunately, it is not easy to disentangle the various effects operating in real solids. It has been suggested that the Anderson mechanism is responsible for metal–insulator transitions in some oxides, such as Na_xWO_3 and $La_{1-x}Sr_xVO_3$, but there are undoubtedly other interactions also at work, such as small polaron formation. The clearest example of an Anderson transition is found in some heavily doped semiconductors. As discussed in Section 7.2.5 these are metallic if the dopant concentration is sufficiently high. We can imagine that there is an impurity band overlapping with the conduction band in a heavily doped n-type semiconductor. Since the impurity band comes from the disordered arrangement of donor atoms, states near the bottom may be localized, with a mobility edge as in Fig. 7.15(c). Electrons can be removed from the band, without changing the donor concentration, by *compensating* with a p-type dopant. Some electrons then fall into acceptor orbitals, and the Fermi level in the conduction band is lowered. At a certain level of compensation it is found that the metallic conductivity ceases, and is replaced by behaviour characteristic of localized electrons. This is believed to be an Anderson transition, happening when the Fermi level falls below the mobility edge.

7.5 Excited states

Electronic excited states are produced in solids by absorption of photons of sufficient energy, or by bombardment with charged particles such as electrons. In non-metallic solids, free electrons and holes may be formed if the energy of the radiation exceeds the band gap. However, the threshold of the absorption spectrum may be complicated by the appearance of **excitons**, where the electron and hole are still held together by their electrostatic attraction. Excited states may be regarded as **non-equilibrium defects**; it is interesting to look at them in more detail at this point, especially as many of their properties are the result of interaction with other defects.

7.5.1 Excitons

An electron in the conduction band is attracted to a hole in the valence band by an electrostatic force, similar to that between an electron and a positively charged lattice defect. Thus a bound state can be formed, in which the electron and hole are held together by their attraction. The strength of the binding is determined by the same parameters as in a donor or acceptor state introduced by a charged defect (see Section 7.2.1), viz.:

1. The dielectric constant of the medium: the higher the value of this, the weaker the attraction.
2. The effective masses of the electron and hole: the smaller these are, the more difficult it is to hold the pair together.

In semiconductors with wide bands and small gaps, m^* is small and ε_r large, and only very weak binding occurs. We can imagine that the electron and hole 'orbit' each other in a state with a large radius (see Fig. 7.16). This weakly bound state is known as a **Wannier exciton**. The energy is given by a hydrogenic formula similar to that for defect states: in the case of an exciton the unbound state corresponds to the free electron and hole, with an energy

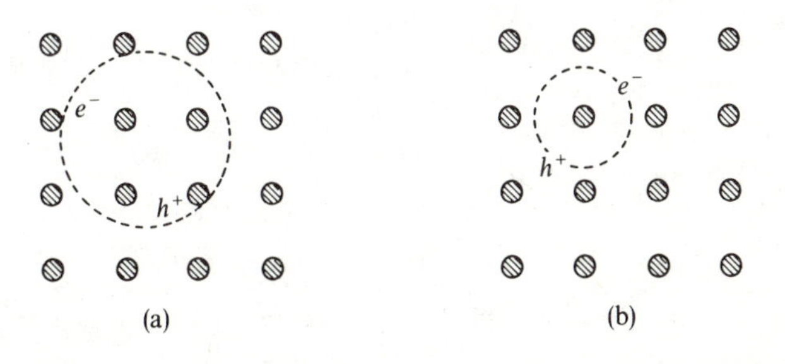

Fig. 7.16 Excitons. (a) Weak binding of electron and hole gives an extended Wannier exciton. (b) Strongly bound Frenkel exciton, with electron and hole on the same centre.

equal to the band gap, E_g. Thus the exciton has energies:

$$E_n = E_g - \mu e^4/(32\pi^2\varepsilon_0^2\varepsilon_r^2\hbar^2 n^2) \tag{7.26}$$

where μ is the reduced mass of the electron–hole pair:

$$\mu = m_e m_h/(m_e + m_h). \tag{7.27}$$

$n = 1$ in equation 7.26 corresponds to the most strongly bound state (like a 1*s* orbital), but higher energy states with $n = 2, 3, \ldots$, can sometimes be seen.

Wannier excitons are not seen in the absorption of silicon and germanium, because these solids have indirect band gaps, so that the lowest energy excitation is forbidden, and only ocurs in conjunction with lattice vibrations. Many compound semiconductors such as GaAs have direct gaps, however, and exciton peaks can be seen at the threshold of the absorption spectrum. The optical absorption of GaAs was shown in Fig. 2.11, on p. 37. The clearest example of a Wannier exciton is found in cuprous oxide (Cu_2O) the spectrum of which is shown in Fig. 7.17. The band gap of this compound is 2.16 eV, and just below this energy can be seen a series of peaks corresponding to successive values of *n* in equation 7.26. Exciton structure such as this can generally only be seen at rather low temperatures. Lattice vibrations present at higher temperatures interact with the electron and hole and cause the exciton to split up. Thus

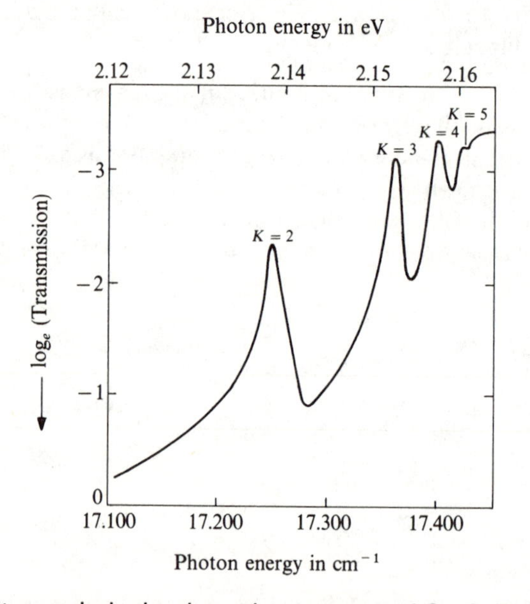

Fig. 7.17 Exciton peaks in the absorption spectrum of Cu_2O. Transitions to the n = 1 level in this solid are forbidden by symmetry. (After P. W. Baumeister, *Phys. Rev.*, **121** (1961), 359.)

the exciton peaks in the spectrum broaden and merge into the edge of the band-to-band excitation.

In materials with narrower bands and lower dielectric constants, the binding between electrons and holes is stronger, and the wave function of the exciton more compact. The limit of the tightly-bound situation is called a **Frenkel exciton**, and corresponds to the electron and hole held together on the same atom or molecule. This is also shown in Fig. 7.16. Frenkel excitons and found in many molecular crystals, for example in the π to π^* transitions in aromatic molecules such as anthracene. Other examples are the ligand field excitations of localized *d* electrons, such as in nickel oxide. There is an interesting connection between the formation of excitons and the Hubbard model for electron localization, explained in Section 5.1. We have been using the language of electrons and holes, but of course a hole is a fictitious particle, representing the absence of an electron in an occupied orbital. The attraction of the electron and hole, which binds them in a Frenkel exciton, is another way of describing the extra repulsion that the electron would experience if it moved to another site, where there was no hole. This extra electron repulsion is the Hubbard U responsible for electron localization in compounds such as NiO. Thus the Frenkel exciton is formed in the excited state by the same interaction that gives localized *d* electrons in the ground state.

Although the electron and hole in a Frenkel exciton are held together on the same lattice site, the exciton as a whole may move through the crystal, by transferring its energy to a neighbouring site. This can be seen in crystals containing impurities which can trap the excitation. For example, pure anthracene shows fluorescence when it is excited optically. If it is doped with a few parts per million of tetracene, the fluorescence from the anthracene diminishes, but appears instead from the impurity. This shows that tetracene acts as an efficient trap for the exciton, and the low concentrations involved imply that the excitons can travel over a hundred molecules or more before de-excitation. The interaction causing the exciton to move may sometimes come from overlap of orbitals on adjacent sites. With allowed transitions such as π to π^* however, the most important term is the electrostatic interaction of the transition dipoles:

$$\mu_{if} = \int \Psi_i \mu \Psi_f \tag{7.28}$$

where Ψ_i and Ψ_f are the wave functions for the ground and excited states. The dipole–dipole interaction gives a coupling between excitations:

$$t \propto \mu_{if}^2 / R^3 \tag{7.29}$$

The interaction between neighbouring molecules in their excited state can have more subtle effects than the simple migration of excitons. A wave function in which an identifiable molecule is excited is not a correct eigenfunction of the crystal as a whole. Instead we must write linear combinations of excited states, like the combinations of atomic orbitals used in band theory. In a one-dimensional case, with one molecule per unit cell and a lattice spacing a, we can write:

$$\Psi_k = \sum_n \exp(ikna)\phi_n. \qquad (7.30)$$

In this equation ϕ_n refers to the wave function where the molecule at site n is excited, and all others are in their ground state. It is important to realize that the state Ψ_k has only *one* excited state present, although it is no longer confined to one site, but is in some sense delocalized through the crystal. Using the interaction t between adjacent excitations, we can calculate the energy of the state Ψ_k, just as in the LCAO theory:

$$E(k) = E_0 + 2t\cos(ka). \qquad (7.31)$$

E_0 is the energy of an isolated excited state, but as a result of the interaction there is a whole band of exciton energies in the crystal. Normally we do not see the width of this band, because the optical selection rule $\Delta k = 0$ means that we can only produce the exciton with $k = 0$, where the excitation on all molecules is in phase. This excitation is shifted in energy, however, by $2t$ with respect to the isolated molecule. This so-called **Davidov shift** is nicely illustrated in the charge-transfer bands of compounds containing the tetracyanoplatinate ion $[Pt(CN)_4]^{2-}$. By changing the counter cations in the lattice, it is possible to get a range of Pt–Pt distances, and Fig. 7.18(a) shows that the energy of the excitation varies linearly with R^{-3}. This is expected from the dipole–dipole formula, equation 7.29. Extrapolation of the line to infinite distance predicts a transition energy of 44 800 cm^{-1}, which is close to the value of 46 000 cm^{-1} found for the solution spectrum of isolated $[Pt(CN)_4]^{2-}$ ions.

More complex effects occur when the unit cell of the crystal contains two or more molecules. Like the case of the binary chain in band theory (see Section 4.1.3 on p. 88), each function Ψ_k is now a linear combination of the different excitations possible within one unit cell. For example, with two molecules, we can write:

$$\Psi_k = \sum_n \exp(ikna)\,[a_n\phi_n^{\mathrm{a}} + b_n\phi_n^{\mathrm{b}}] \qquad (7.32)$$

where ϕ_n^{a} and ϕ_n^{b} are functions in which molecules a and b at site n are excited. For each value of k, there are two possible combinations, with different values of a_n and b_n. In particular, for the $k = 0$ exciton seen in the optical spectrum, there are states with two different energies. Thus the interaction between the two molecules in the unit cell produces a **Davidov splitting** of the absorption line. This can be seen in the absorption spectrum of solid anthracene, where there are two molecules per unit cell, in different orientations. Figure 7.18(b) shows that a single peak in the solution spectrum is split into two in the solid.

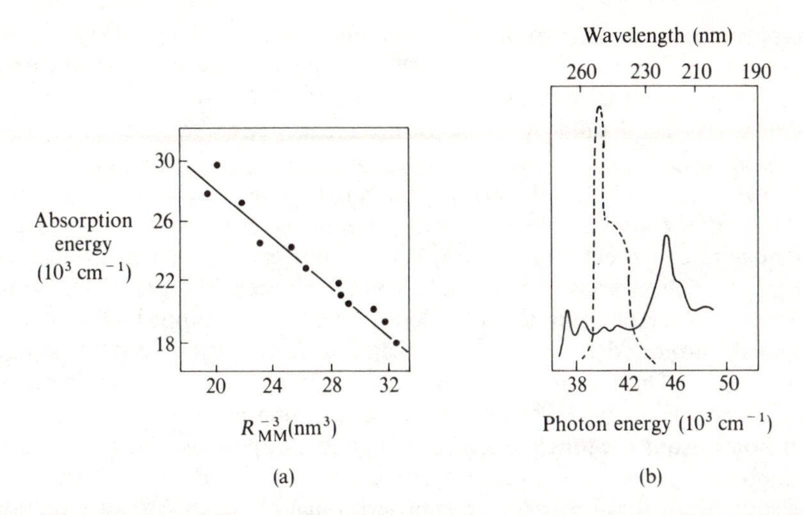

Fig. 7.18 Davidov shifts and splittings. (a) Absorption energies of $[Pt(CN)_4]^{2-}$ in various solids, plotted against inverse cube of Pt–Pt distance. (After P. Day, *J. Am. Chem. Soc.*, **97** (1975), 1588.) (b) Spectrum of solid anthracene (———) shows splitting of single peak observed in the solution spectrum (– – – –). (After L. E. Lyons and G. C. Morris, *J. Chem. Soc.* (1959), 1551.)

7.5.2 Trapping of electrons and holes: luminescence

It was mentioned in the previous section that excitons can be trapped by impurities, which sometimes produce their own fluorescence. Free electrons and holes may also be trapped in this way. When the solid is excited by an electron beam, this gives the phenomenon known as **cathodiluminescence**, which is used in cathode ray tubes (CRTs). The traditional phosphor for CRT screens is zinc sulphide doped with transition-metal impurities such as manganese, but modern phosphors tend to be based on complex oxides such as yttrium aluminium garnet, $Y_3Al_5O_{12}$ (YAG), doped with transition metals or lanthanides. The high energy electron beam loses energy by producing secondary electrons and holes in the host lattice. These in turn are trapped by the dopant ion, leaving it in an electronic excited state, which emits radiation as it decays to the ground state.

There is some evidence that the first stage in the trapping of electrons and holes in a cathodiluminescent material may be a redox process in which the oxidation state of the dopant ion is changed. For example, in lanthanide-doped YAG the trapping is found to be more efficient with Ln^{3+} ions that can be oxidized to Ln^{4+} (cerium and terbium) or reduced to the Ln^{2+} state (europium). In the former case, the lanthanide ion probably acts by first trapping a hole from the valence band, thus being oxidized. This is one of the processes illustrated in Fig. 7.2 (p. 198). The oxidizable Ln^{3+} has an occupied level just above the valence band, thus acting as a trap for holes (Fig. 7.2(f)). Once the Ln^{4+} state has been formed, the extra positive charge can then attract and trap an electron, leaving the Ln^{3+} in an excited state which luminesces. In a similar way, Eu^{3+} can first trap an electron to form Eu^{2+}, so that it acts as an unoccupied level just below the conduction band, as in Fig. 7.2(e).

Sometimes electrons and holes may be trapped at defect sites a long distance apart, and the probability of recombination, by either radiative or non-radiative means, is extremely low. If the trapping energy is large enough, the lifetime of such an 'excited state' can be very long indeed. Many minerals in fact contain appreciable numbers of electrons and holes trapped in this way, originally produced by background radiation from cosmic rays or natural radioactivity. These solids often show the phenomenon of **thermoluminescence**, since by heating to temperatures of a few hundred degrees, the electrons and holes acquire enough thermal energy to escape from their traps and recombine. Thermoluminescence is used as a method for dating pottery, since the strength of the luminescence is proportional to the dosage of radiation received. When the pot is first fired, all the electrons and holes are released from their traps, so that the 'clock' is reset. The thermoluminescence measured in the laboratory is proportional to the total background radiation received by the pot after this time. Once the measurement has been made, the sensitivity of the particular material can be calibrated by irradiating under laboratory conditions. To find the age, it is then necessary to estimate the strength of the different sources of radiation at the point where the pot has been stored or buried.

7.5.3 Chemical reactions of electrons and holes

Photochemical reactions, produced by electronic excited states, are well known in gas-phase and solution chemistry. Many solids also show such reactions, and some of these are of great technological importance, for example the slow deterioration of many polymers on exposure to sunlight. One of the most important and extensively investigated solid-state photochemical reactions is the reduction of silver halides, used in photography. The photoreduction of silver halides causes a darkening due to small clusters of elemental silver. Under some circumstances, for example when very small crystals are incorporated into an oxide glass, the darkening is reversible, and this is used in photochromic lenses, which darken in bright sunlight, and become clearer in less intense light. In a short photographic exposure however, the darkening is not immediately visible, but enough silver particles are produced to act as catalytic centres for the subsequent reduction of the halide by the developer.

At room temperature, silver bromide has a significant concentration of Frenkel defects. The interstitial silver ions give rise to empty defect levels close to the conduction band, and these act as traps for the electrons produced by light. After exposure to band-gap radiation, a far infra-red absorption band can be seen in AgBr. It is believed that the energy of this (167 cm^{-1} or 0.02 eV) is that required to ionize the electrons from the traps. The interstitial silver atoms are quite mobile in the lattice, and some may migrate to the surface of the crystal, where they form even more stable traps. The following series of reactions (in which Ag_i^+ represents an interstitial silver ion) leads to the formation of small silver clusters:

$$Ag_i^+ + e^- \rightarrow Ag$$
$$Ag + e^- \rightarrow Ag^-$$
$$Ag^- + Ag_i^+ \rightarrow Ag_2$$
$$Ag_2 + Ag_i^+ \rightarrow Ag_3^+$$
$$Ag_3^+ + e^- \rightarrow Ag_3 \rightarrow Ag_3^- \rightarrow \ldots.$$

The silver clusters on the surface of the AgBr crystal form the **latent image**, and act as a catalyst for further reduction in the development process. It is essential to the photographic process that electrons and holes produced by the radiation separate without recombination, and the band structure of silver bromide, discussed briefly in Section 4.3.1, contains two features which help this:

1. The $E(k)$ function curves more strongly in the conduction band than in the valence band, so that electrons have a lower effective mass than holes. The electrons in consequence are very mobile, whereas the holes are easily trapped by lattice distortions, and have very low mobility.
2. There is an indirect band gap, so that direct radiative recombination of electrons and holes is forbidden.

The band gap of silver bromide is 2.7 eV, which means that only visible light in the extreme blue end of the spectrum can be directly absorbed. The spectral sensitivity of photographic emulsions is normally extended to lower energies by the use of **sensitizers**. These are organic dye molecules which absorb lower photon energies than AgBr itself. Although the mechanism of sensitization is not completely certain, it seems that in most cases it involves the transfer of electrons from the excited state of the dye into the conduction band of the silver halide. This is shown in Fig. 7.19. The injected electrons can act in the same way as those produced by absorption in the halide itself.

7.6 Surfaces

The electronic structure of a surface is important chemically as it may control the way in which molecules adsorb and react in heterogeneous reactions. Many electronic properties of materials, such as those involved in the emission of electrons, and in electrochemical reactions, also depend on surface effects. For

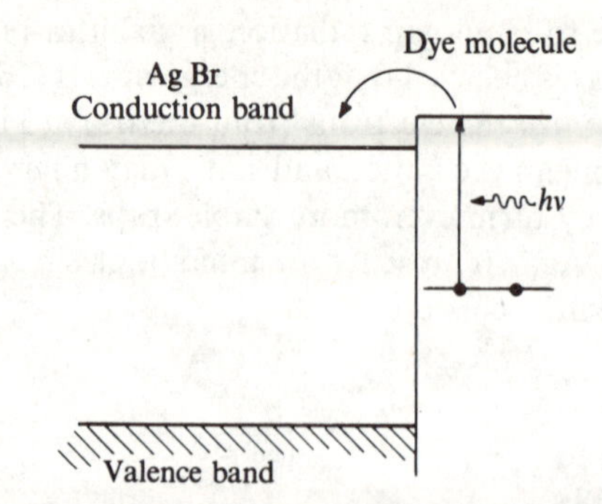

Fig. 7.19 A mechanism for the sensitization of the photographic process in silver halides. An organic molecule adsorbed on the halide surface has an excited state above the energy of the conduction band, and so after absorption of light, can transfer an electron into the solid.

these reasons surfaces have been the subject of intensive research in recent years. A full account of surface electronic structure would require a large volume to itself; here, only a look at a few of the most important aspects is possible.

7.6.1 The work function

The normal reference point for energies within a solid is the Fermi level, E_F. Outside the solid, however, the natural zero of energy is the **vacuum level,** E_V, which is the energy of an electron at rest, and far away from any electrical charge. The relation between the two energies is given by the **work function**, defined by:

$$\Phi = E_V - E_F. \tag{7.33}$$

The work function of a metal is therefore the minimum energy required to remove an electron from the solid into the vacuum. This is shown in Fig. 7.20.

Work functions for metals are found to be in the range 2–5 eV. As might be expected, there is an approximate correlation with the atomic ionization energy, the lowest value (1.9 eV) for any element being that of caesium. However, an actual calculation of work functions is very difficult. Not only is it necessary to know the energy of electrons in the bulk of the solid, but there are also various surface effects, which can give rise to slightly different work functions for different crystal faces of the same solid. Quite large changes of work function may be produced by adsorption of foreign atoms onto a surface, and these are exploited for different purposes. For example, in a photomultiplier or image-intensifier device, in-coming photons produce electrons from a metal or semiconductor surface. The work function represents the minimum photon energy that can be detected in this way. It has been found

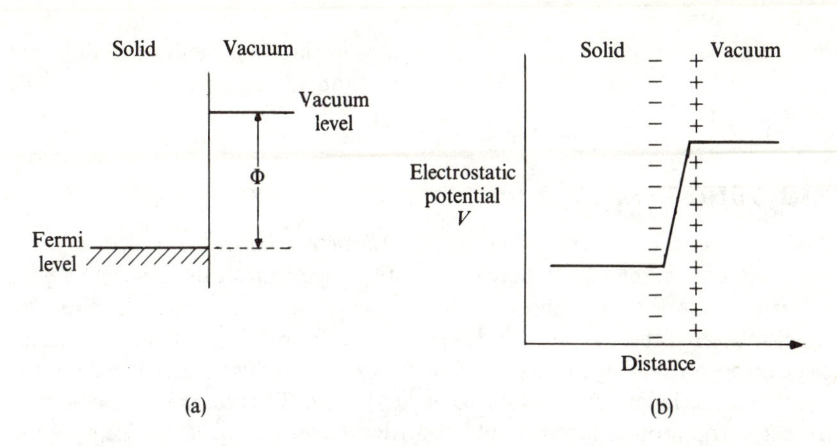

Fig. 7.20(a) The work function is the difference between the Fermi level and the vacuum level. (b) The variation of electrostatic potential produced by a dipolar layer at the surface. A potential V contributes a term $-eV$ to the electron energy, and hence to the work function.

that coatings of caesium, laid down on a semiconductor in the presence of oxygen, can give surfaces with work functions as low as 1 eV, corresponding to photons in the near infra-red. It is now known that these coatings contain sub-oxides of caesium, such as $Cs_{11}O_3$. The detailed reasons for the low work function are not known; however, one way in which a surface layer can influence the work function is shown in Fig. 7.20(b). If there is some charge transfer between an adsorbate and the surface, this can produce a dipolar layer, which alters the electrostatic potential felt by an electron as it passes through the surface. In the case shown, the positive end of the dipoles point outwards. The change in potential through the surface then makes it easier to move an electron out into the vacuum, and so lowers the work function.

The work function is also important in **thermionic emission** from solids, that is the emission of thermally excited electrons from a hot surface. Thermionic emitters are used in electron guns, for example in television tubes. The maximum current density that can be produced at a given temperature is given by the **Richardson equation**:

$$i = A\,T^2 \exp(-\Phi/kT). \tag{7.34}$$

A is roughly the same for all metals, and it can be seen from this equation that at any temperature, solids with the lowest work functions will give the maximum emission. The most efficient thermionic emitters contain oxide coatings, for example BaO on a refractory metal such as osmium. There is some evidence that the oxide layer forms with a metal atom outwards from the surface, so that the work function is lowered by a surface dipole, as in Fig. 7.20(b). Another good emitter is the metallic compound lanthanum hexaboride, LaB_6. The electronic structure of hexaborides was mentioned briefly in Section 5.2.2; the reason for the low function of these solids is not fully understood.

7.6.2 Surface electronic states

Since the atoms on a surface are in a different environment from those of the bulk solid, we might expect there to be distinct electronic levels associated with a surface, similar to those produced by defects. The most important experimental technique for studying surface electronic structure is photoelectron spectroscopy. For electrons with energies in the range 10–100 eV, the effective path length in the solid is around 1 nm. This means that electrons from the top atomic layer should contribute around 10–20 per cent of the photoelectron spectrum. Unfortunately, it is not always easy to separate the signals from bulk and surface atoms, but various methods can be used to help in this. For example, if the photoelectrons are observed coming at a nearly glancing angle from the solid, the path length from the bulk is increased, and so the surface signal should be enhanced. It is also possible to study the effect of chemically modifying the surface, so as to create or remove surface states in a controlled way.

In ionic solids, the importance of the Madelung potential for determining the energies of the valence and conduction bands was emphasized in Section 3.1.1. Atoms at the surface have a lower coordination, and the Madelung potential should be reduced from the bulk value. This effect on its own should make the surface band gap less than in the bulk, and so give surface levels at energies within the bulk band gap. Experimentally, however, there is no good evidence for occupied band gap states on flat, defect-free surfaces of ionic solids. There may be several reasons for this. In the first place, the most stable surfaces (such as the (100) cubic faces common with the NaCl structure) are those showing the least perturbation from the bulk structure, so that the change in Madelung potential is small. It was also pointed out previously that although the Madelung term is important, it is not the only factor determining the band energies. There are the bandwidths to consider, and the polarization energies involved in creating charges. These effects will also be less on the surface, and will give a contribution to the surface shift in the opposite direction to that of the Madelung potential. Another important factor is that atoms at the surface may relax from their ideal bulk positions. As in the case of semiconductors discussed below, this relaxation may have the effect of stabilizing filled levels.

On surfaces of covalent solids, the reduced coordination would be expected to lead to dangling bonds, like those found in amorphous silicon (see Section 7.4.1). Being non-bonding levels, dangling bonds should also produce states within the bulk band gap. Once again, the experimental situation is more complicated. Numerous structural studies have shown that many semiconductor surfaces, especially those expected to have dangling bond states, are extensively relaxed, or even *reconstructed* from their ideal bulk positions. This is not surprizing, since large numbers of dangling bonds would be highly unfavourable, and the surface atoms might be expected to move to positions where they could achieve greater bonding stability. The precise details of many of the reconstructions, for example on the surface of silicon, are still controversial. One example of a surface relaxation that is quite well understood is shown in Fig. 7.21. The (110) surface of Group III–V semiconductors such as GaAs should have two states within the band gap; these are essentially dangling bond states associated with the two kinds of atom. It is found that the relaxation of the surface causes the Group V atoms move outwards relative to the others. Calculations show that this stabilizes the lower levels, and destabilizes the upper ones. Although surface states are still found within the band gap, they are considerably closer to the band edges than in the unrelaxed surface.

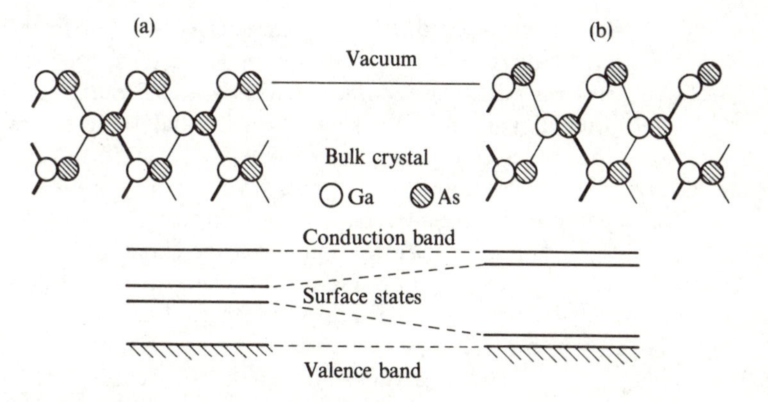

Fig. 7.21 Structure and electronic levels of a (110) surface of GaAs. (a) Unrelaxed surface, with predicted states well within the band gap. (b) Observed relaxation of surface, and effect on electronic levels.

These examples show that the changes of electronic structure at a clean, flat surface are generally less marked than would be expected from simple ideas. More serious changes can occur however, when the surface is very rough, or has defects present on it. Figure 7.22 shows the photoelectron spectrum of strontium titanate after the surface has been etched with high-energy argon ions. The spectrum shown previously in Fig. 7.13 has virtually no signals from the band-gap region between the Fermi level and the valence band edge. The spectrum after etching shows that there are now occupied electronic states within the gap. The etching process removes oxygen more easily than it does

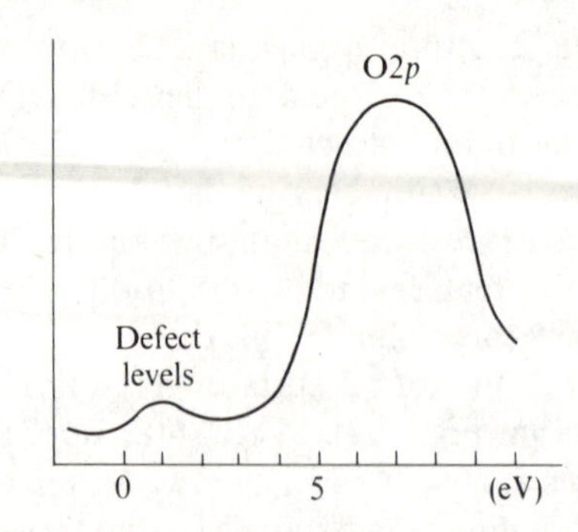

Fig. 7.22 Photoelectron spectrum of $SrTiO_3$ after etching with argon ions. The band-gap surface defect levels are marked. (Compare with Fig. 7.13(a)).

heavier elements, and leads to a reduction of some surface titanium from Ti^{4+} to Ti^{3+}. The surface states containing the titanium 3*d* electrons are probably associated with oxygen vacancies. Surface defects such as these are quite important, since they may often be the sites at which catalytic reactions and other interesting processes take place.

Extra electronic levels also appear on surfaces when atoms or molecules are adsorbed. A comparison of these adsorbate levels with the orbitals of the free species can often give important information about the chemical bonding to the surface. Figure 7.23 shows for example the difference that appears in the photoelectron spectrum when water is adsorbed on rutile, TiO_2. There are three extra peaks, which are compared (see Fig. 7.23) with the spectrum of water in the gas phase. All the orbitals are shifted in the adsorbate spectra, both because of the difference of reference level (the Fermi level in the solid, as opposed to the vacuum level in the gas phase), and because of the polarization produced in the solid by the ionization of an adsorbed molecule. However, the middle peak, which comes from the $3a_1$ orbital of water, shows an increase of binding energy relative to the other two. The stabilization of this orbital shows that it is the one most involved in bonding to the surface. It is thought that the water coordinates to a metal atom, and $3a_1$ orbital is the lone pair orbital that overlaps most effectively with empty *d* orbitals of surface titanium atoms. Another system that has been extensively studied in this way is that of carbon monoxide adsorbed on transition metals. The mode of bonding seems to be similar to that in metal carbonyls: that is through overlap of the carbon lone-pair with empty orbitals at the metal surface, and some 'back-donation' from overlap of filled metal *d* orbitals with the antibonding π orbitals of the molecule.

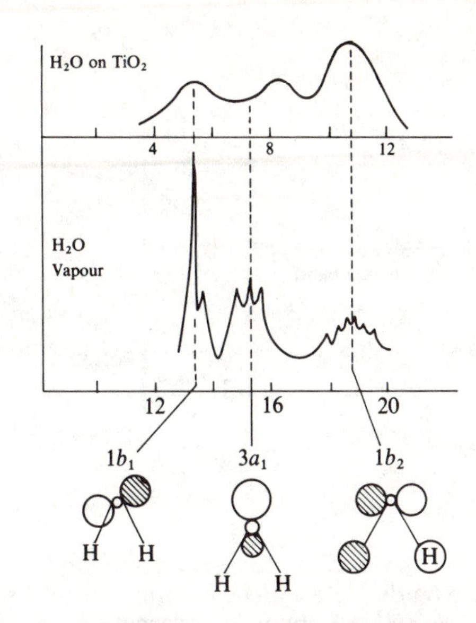

Fig. 7.23 Difference in photoelectron spectrum resulting from adsorption of water on rutile, TiO_2. Comparison with gas-phase spectrum of water, with a rough picture of the molecular orbitals corresponding to each peak.

7.6.3 Band bending at semiconductor surfaces

We have seen that in the bulk of a semiconductor, the position of the Fermi level is usually controlled by the energy levels introduced by doping. At the surface it is the energies of surface states associated with defects or adsorbates that are important, and the position of the Fermi level relative to the band edges may not be the same as in the bulk. Since however the Fermi level is the same throughout a solid in equilibrium, it must be the band energies themselves that are different at the surface. Figure 7.24 shows an example of the **band-bending** that can occur when an n-type semiconductor has a surface state in the band gap. Electrons are transferred from the donor levels of the surface region into the surface states, to give a **depletion layer**, where there is an electrostatic field from the unbalanced positive charge of the ionized donors. It is this field that causes the shift in the band energies. The following simple calculation shows how the degree of band bending depends on the amount of charge transferred to the surface.

The variation of electrostatic potential V is given in terms of the charge density ρ and the relative dielectric constant ε_r of the medium by Poisson's equation:

$$d^2V/dx^2 = \rho/(\varepsilon_0\varepsilon_r). \tag{7.35}$$

Within the depletion layer of an n-type semiconductor, the charge density is

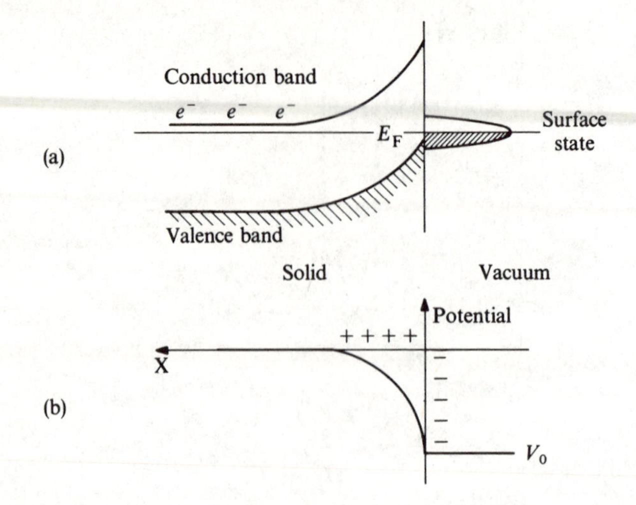

Fig. 7.24 Band bending due to a surface state on a n-type semiconductor. (a) Band energies and Fermi level, showing surface state. (b) Charge distribution and variation of electrostatic potential.

produced by the ionized impurities, so that:

$$\rho = n_d e, \tag{7.36}$$

where n_d is the dopant concentration. Since equation 7.35 is a second-order differential equation, two boundary conditions are required. These are determined (see Fig. 7.24(b)) by measuring the zero of potential from that in the bulk of the semiconductor, and by assuming that there is no electric field away from the depletion layer. Thus:

$$V = 0; \quad dV/dx = 0, \quad \text{for } x > w, \tag{7.37}$$

where w is the width of the layer. It is then easy to see that a solution of Poisson's equation in the depletion layer gives:

$$V = n_d e/(2\varepsilon_0 \varepsilon_r)(w - x)^2 \quad 0 < x < w. \tag{7.38}$$

The total band bending is the potential at the surface, $x = 0$:

$$V_0 = n_d e/(2\varepsilon_0 \varepsilon_r) w^2 \tag{7.39}$$

so that the width of the layer is given by:

$$w = [2\varepsilon_0 \varepsilon_r V_0/(n_d e)]^{1/2} \tag{7.40}$$

The total number of electrons transferred out of the semiconductor is just the number of ionized donors within the depletion layer. This gives a surface charge (q_s) per unit area equal to:

$$q_s = wn_d e = [2\varepsilon_0\varepsilon_r V_0 n_d e]^{1/2} \tag{7.41}$$

These equations show how the amount of band bending, and the width of the carrier-free depletion layer, can be controlled on a given semiconductor by varying the amount of charge taken to the surface. For a semiconductor with a dielectric constant of 10, and a doping level of 10^{17} per cm^3, a band bending of 1 V corresponds to a depletion layer about 100 nm thick, and a surface charge density of around 10^{12} electrons per cm^2. Since the number of surface atoms is of the order of 10^{16} per cm^2, it is clear that only quite a small number of surface states, for example those associated with defects, is sufficient to give an appreciable band bending.

One effect of the depletion layer is to give a high surface resistance. Conductivity measurements on powdered samples are often controlled by surface contacts between the grains, and this can give very misleading results, if a bulk conductivity is required. The occupancy of surface states, and hence the width of the depletion layer, can be altered by the adsorption of molecules from the gas phase. This is used in some solid-state sensors, where the presence of different gases can alter the conductivity. It is found, for example, that the surface resistance of some n-type oxides such as SnO_2 is lowered in air by exposure to reducing gases such as methane. It is thought the adsorption of oxygen removes electrons from the surface, to give species such as superoxide, O_2^-, and hence produces a depletion layer. Molecules that can be catalytically oxidized on the surface then liberate electrons, and hence reduce the band bending and increase the surface conductance.

In many semiconductor applications, it is the *interface* between the semiconductor and some other medium that is important. Figure 7.25 shows a semiconductor–metal interface, where the Fermi level at the interface is *pinned*

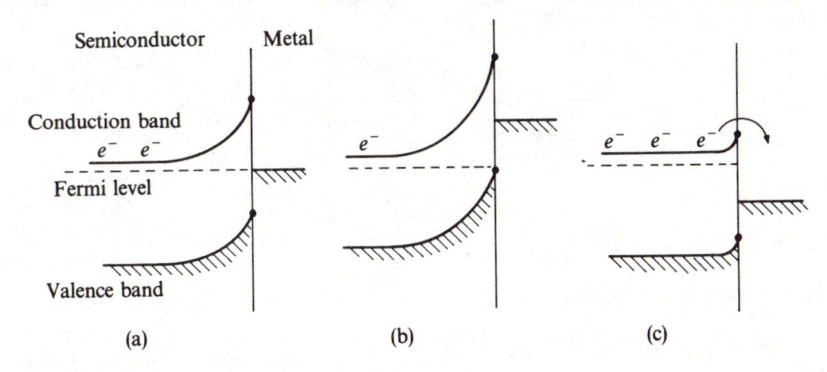

Fig. 7.25 Schottky barrier formed by metal–semiconductor junction. (a) Equilibrium. (b) Reverse bias: no current flows. (c) Forward bias decreases barrier to current flow.

somewhere in the middle of the gap, for example by interface states resulting from a chemical reaction between the two solids. This situation is known as a **Schottky barrier**: the depletion layer gives rise to rectifying properties similar to those in a p–n junction. Figure 7.25 also shows that the band bending is increased by a reverse bias, when the n-type semiconductor is made positive. Easy conduction occurs only under forward bias, which reduces the energy barrier required for electrons to flow from the semiconductor to the metal. Schottky diodes based on this effect have a faster response than ones with p–n junctions, and are often used for high-frequency applications.

In many semiconductor devices, an insulating layer of SiO_2 is produced between the semiconductor and a metal. The number of surface states at a Si–SiO_2 interface is very small, and so the Fermi level in the silicon can be altered without the pinning effect of interface states. One of the most important switching devices in integrated circuits is the **MOSFET**: the metal–oxide–semiconductor-based field-effect transistor. As shown in Fig. 7.26, two regions of p-type silicon, forming the *source* and the *drain*, are separated by n-type material. Normally current between the source and drain is blocked, since one of the p–n junctions must be under reverse bias. However, when a negative potential is applied to the insulated metal *gate*, the bands at the surface are bent up, just as in Fig. 7.24. For silicon, with a band gap of 1.1 eV, a bending of 1 eV is sufficient to push the valence band edge up close to the Fermi level, so as to produce a surface *inversion layer* with some p-type carriers. This allows current to flow between the p-type regions, and so switches on the MOSFET. As shown above, only a small surface charge is required to produce a significant band bending, and it is possible to couple a field-effect transistor to various chemical reactions that involve a charge separation, and so to make chemical sensors.

The final example to consider is the electrochemical interface between a semiconducting electrode and an electrolyte solution. Often it is found that the

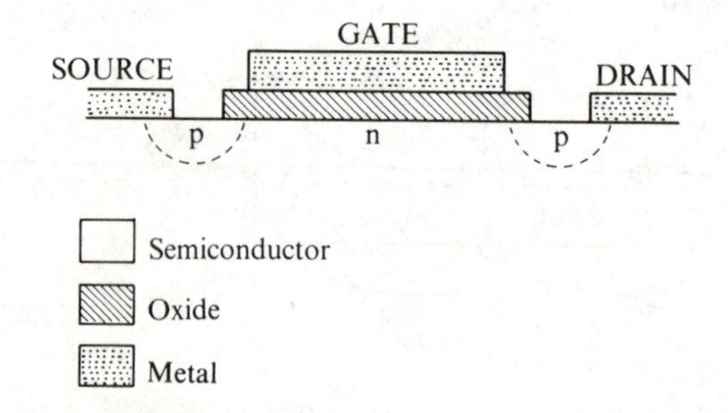

Fig. 7.26 A MOSFET. Current flow between the source and the drain is normally inhibited by the p–n–p junction configuration. Conduction becomes possible when a negative potential applied to the gate causes a sufficient band bending to make a p-type pathway.

position of the semiconductor band edges at the interface are fixed, as in a Schottky barrier. The surface Fermi level position, and hence the band bending, may then be altered by varying the potential of the semiconductor with respect to some reference electrode. The potential where no band bending is present is known as the **flat-band potential**, and is important as it gives the energies of the semiconductor valence and conduction bands relative to those of redox processes in solution. In simple cases, the flat-band potential can be determined by measuring the electrical capacitance of the depletion layer, as a function of potential. The amount of charge removed from the layer is given by equation 7.41. The capacitance measured in an alternating current circuit is the differential of this with respect to potential. Thus:

$$C = \mathrm{d}q/\mathrm{d}V = [\varepsilon_0 \varepsilon_r n_d e/(2V - 2V_{fb})]^{1/2} \tag{7.42}$$

where V is the applied potential, and V_{fb} the flat-band potential. A *Mott–Schottky plot* of $1/C^2$ against V should have an intercept at the flat-band potential.

Band bending at the semiconductor–electrolyte interface is important in **photoelectrochemistry**. Section 7.2.4 described the operation of a simple solar cell, where electrons and holes generated by absorption of light are separated by the band bending in a p–n junction. The same effect is produced by the band-bending present at a surface, and is illustrated in Fig. 7.27. In an n-type semiconductor, electrons will migrate into the bulk, and holes to the surface, where they may produce electrochemical reactions, thus acting as a **photoanode**. A p-type semiconductor may similarly act as a **photocathode**, with electrons coming to the surface. In Fig. 7.27, the two semiconductor electrodes form parts of the same cell; it is also possible to use a metal such as platinum as a counter-electrode. When opposite redox processes take place at the two electrodes, there is no net chemical reaction, but a potential can be generated, as in a photovoltaic device. For example, a solar to electrical energy conversion

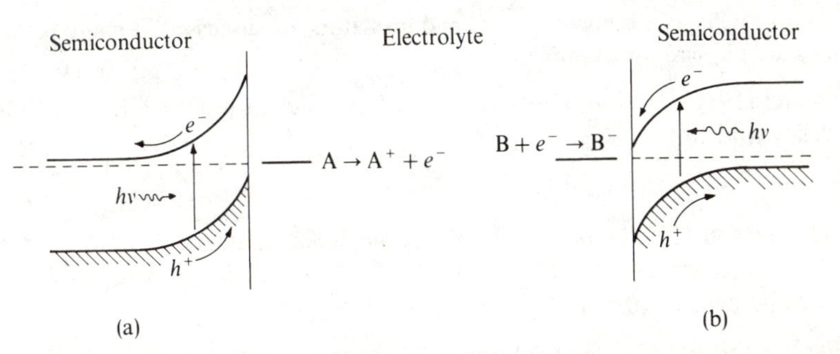

Fig. 7.27 Semiconductor photoelectrodes. (a) An n-type photoanode. (b) A p-type photocathode. Electrons and holes produced by light absorption are separated by the surface band bending.

efficiency of 11.5 per cent has been achieved in a cell where a p-type InP photocathode reduces V^{3+} in aqueous solution:

$$V^{3+} + e^- \rightarrow V^{2+}$$

The corresponding re-oxidation occurs at a platinum counter-electrode. Various problems have to be overcome before a useful efficiency can be obtained. n-type semiconductors acting as photoanodes are very susceptible to corrosion, because the holes can be trapped at the semiconductor surface to produce oxidation, for example of S^{2-} to sulphur, which inhibits the action. It is also essential to remove surface states that can act as traps where carriers recombine and therefore 'short-circuit' the desired cell reactions.

An attractive alternative to a photovoltaic device is one using an electrochemical reaction to generate a fuel such as hydrogen. The electrolytic splitting of water requires a free energy input of 1.23 eV per electron, and theoretically any semiconductor with a band gap greater than this should be usable in a photoelectrolytic cell. However, there are many losses involved, and the unassisted photoelectrolysis of water has only been achieved so far with wide band gap materials such as n-type TiO_2 and $SrTiO_3$. They will act as photoanodes, and can evolve oxygen when illuminated with band gap radiation, the hydrogen being evolved at a metal counter electrode. Unfortunately, the band gap of 3 eV in these oxides is just outside the visible region of the spectrum, and only a very small fraction of solar radiation reaching the earth's surface is absorbed. Thus the overall efficiency of conversion is very low.

Further reading

Aspects of defects in semiconductors and insulators are discussed in many books on solid state physics, for example in:

C. Kittel (1976). *Introduction to solid state physics* (5th edn), Chapter 8. John Wiley and Sons.

Two good accounts concentrating on defects and their properties are:

B. Henderson (1972). *Defects in crystalline solids.* Edward Arnold.

A. M. Stoneham and W. Hayes (1985). *Defects and defect processes in non-metallic solids.* John Wiley and Sons.

Electronic theories of defects are treated in particular detail in:

A. M. Stoneham (1985). *Theory of defects in solids* (2nd edn), Oxford University Press.

An account of the applications of doped semiconductors in solid-state devices is:

R. Dalven (1980). *Introduction to applied solid state physics.* Plenum Press.

Electronic properties of defects in oxides are treated in:

D. Adler (1967). In *Treatise on solid state chemistry*, Vol. 2 (ed. N. B. Hannay). Plenum Press.

The photographic process is discussed in:

J. F. Hamilton (1973). *Prog. Solid State Chem.* **8** 167.

Accounts of the electronic properties of disordered solids are:

N. F. Mott and E. A. Davis (1979). *Electronic processes in non-crystalline solids* (2nd edn). Oxford University Press.
S. R. Elliott (1983). *Physics of amorphous materials.* Longman.
J. Robertson (1983). *Adv. Phys.* **32**, 36.

A good general account of the properties of surfaces, with an emphasis on catalysis, rather than electronic structure, is:

G. A. Somorjai (1981). *Chemistry in two dimensions: Surfaces.* Cornell University Press.

The following reviews are particularly relevant to the discussion in the present chapter:

V. M. Bermudez (1981). *Prog. Surf. Sci.* **11** 1.
V. E. Henrich (1983). *Prog. Surf. Sci.* **14** 175.
Yu. V. Pleskov (1984). *Prog. Surf. Sci.* **15** 401.

Appendices

These appendices contain material for the slightly more mathematically-inclined reader. Elementary derivations are given of two important results: the **Fermi–Dirac distribution** of electrons at temperatures above absolute zero, and the construction and properties of **Brillouin zones** in a two- or three-dimensional crystal lattice.

A: The Fermi–Dirac distribution function

The normal derivation of the Boltzmann distribution needs to be modified to take account of two properties of electrons:

(i) that they are indistinguishable;

(ii) that they obey the exclusion principle, so that when the spin quantum number is included, only one electron can occupy each state.

Our derivation is based on the **micro-canonical ensemble**. This means that we consider a fixed total number of electrons N, at a fixed total energy U. Imagine a set of levels with energies $\{E_i\}$. Each level has a degeneracy g_i, which shows the maximum number of electrons that it can hold. We shall investigate a distribution in which $\{n_i\}$ are the number of electrons actually in each level. These occupations must give the correct total number of electrons:

$$N = \sum_i n_i \tag{A.1}$$

and the correct total energy, which if the electrons do not interact with each other, is:

$$U = \sum_i n_i E_i. \tag{A.2}$$

The idea, as with the Boltzmann distribution, is to find the most probable distribution. The number of ways that n identical objects can be distributed between g states is simply the binomial coefficient:

$$_gC_n = g!/[n!(g-n)!].$$

Thus the total number of ways in which our particular distribution can be realized is:

$$W = \prod_i g_i!/[n_i!(g_i-n_i)!] \tag{A.3}$$

We use Stirling's approximation in the form:

$$\ln(n!) = n\ln n - n$$

so that:

$$\ln W = \sum_i [g_i \ln g_i - n_i \ln n_i - (g_i - n_i)\ln(g_i - n_i)]. \tag{A.4}$$

Now if the occupation numbers n_i each change by δn_i, we have by differentiating:

$$\delta \ln W = \sum_i [\ln(g_i - n_i) - \ln n_i]\,\delta n_i$$

To find the most probable distribution, we need to maximize W, and hence also $\ln W$, with respect to these variations. Thus:

$$\sum_i [\ln(g_i - n_i) - \ln n_i]\,\delta n_i = 0 \tag{A.5}$$

However, the constraints on the total number of electrons and the total energy must be satisfied, and so by differentiating equations A.1 and A.2, we get:

$$\sum_i \delta n_i = 0 \tag{A.6}$$

and

$$\sum_i E_i \delta n_i = 0 \tag{A.7}$$

These constraints may be incorporated by the method of Lagrangian undetermined multipliers. The three above equations can be satisfied if:

$$\ln(g_i - n_i) - \ln n_i - \alpha - \beta E_i = 0 \tag{A.8}$$

for all energy levels, where α and β are numbers which we need to fix by physical arguments later. From equation A.8:

$$(g_i - n_i)/n_i = \exp(\alpha + \beta E_i)$$

which can be rearranged to give the distribution:

$$n_i/g_i = 1/[1 + \exp(\alpha + \beta E_i)]. \tag{A.9}$$

We shall first show that β is related to the absolute temperature in the same way as in the Boltzmann distribution. The easiest way to do this is to imagine a situation where the number of available levels greatly exceeds the number of electrons, so that the fractional occupancy n_i/g_i of each level is small. This requires that:

$$\alpha + \beta E_i \gg 0$$

or

$$\exp(\alpha + \beta E_i) \gg 1 \quad \text{for all } i.$$

Then we can approximate equation A.9 to give:

$$n_i = g_i e^{-\alpha} \exp(-\beta E_i). \tag{A.10}$$

This is of course the Boltzmann distribution, which we expect to be obeyed under conditions where there are plenty of energy levels to go round, so that the exclusion principle has no effect. We must therefore have:

$$\beta = 1/kT. \tag{A.11}$$

Instead of the other parameter α, we shall define:

$$\mu = -kT\alpha \tag{A.12}$$

and so equation A.9, giving the fractional occupancy, becomes:

$$f_i = n_i/g_i = 1/[1+\exp\{(E_i-\mu)/kT\}]. \tag{A.13}$$

μ is the thermodynamic **chemical potential** for electrons, as may be shown by calculating the free energy of the system. In chemistry we normally work at constant pressure, and use the Gibbs function G. However, in the present case it is easier to assume that we are working at constant volume. This is because the energy levels E_i will normally depend on the volume. We therefore calculate the Helmholtz free energy:

$$A = U - TS. \tag{A.14}$$

The entropy is given by the usual statistical formula:

$$S = k\ln W. \tag{A.15}$$

Substituting the appropriate occupation numbers from equation A.13 into the expression for ln W gives, after some rearrangement:

$$\begin{aligned} S = k\sum_i \{&g_i\ln[1+\exp\{-(E_i-\mu)/kT\}] \\ &+[(E_i-\mu)/kT]/[1+\exp\{(E_i-\mu)/kT]\}. \end{aligned} \tag{A.16}$$

But the second term in this sum is equal to:

$$\begin{aligned} &k\sum_i g_i[(E_i-\mu)/kT]/[1+\exp\{(E_i-\mu)/kT]\} \\ &= k\sum_i n_i(E_i-\mu)/kT \\ &= U/T - \mu N/T \end{aligned}$$

from equations A.13, A.1, and A.2. So:

$$S = U/T - \mu N/T + k\sum_i g_i\ln[1+\exp\{-(E_i-\mu)/kT\}] \tag{A.17}$$

and:

$$A = N\mu - kT\sum_i g_i\ln[1+\exp\{-(E_i-\mu)/kT\}] \tag{A.18}$$

The final step is to differentiate with respect to the total number of electrons N. We need to be a little careful in doing this, since μ will itself change with N. Only the first term depends explicitly on N, so that:

$$(\partial A/\partial N)_{T,V} = \mu + (\partial\mu/\partial N)\left\{N - kT(\partial/\partial\mu)\sum_i g_i \ln[1+\exp\{-(E_i-\mu)/kT\}]\right\}$$

Fortunately, the $(\partial\mu/\partial N)$ terms cancel one another. This is because:

$$(\partial/\partial\mu)\sum_i g_i \ln[1+\exp\{-(E_i-\mu)/kT\}]$$

$$= \sum_i (g_i/kT)\exp\{-(E_i-\mu)/kT\}/[1+\exp\{-(E_i-\mu)/kT\}]$$

$$= (1/kT)\sum_i g_i/[1+\exp\{(E_i-\mu)/kT\}]$$

$$= N/(kT)$$

using equations A.13 and A.1. Thus we find:

$$\mu = (\partial A/\partial N)_{T,V} \tag{A.19}$$

which is the definition of chemical potential, in a system at constant volume. The normal thermodynamic conditions for equilibrium between two phases show that the chemical potentials for species distributed between two phases must be equal. In the theory of solids, the chemical potential μ for electrons is known as the Fermi level, and is written E_F in the text.

B: Brillouin zones and the reciprocal lattice

In the LCAO approach to band theory, which is explained in Chapter 4, the first Brillouin zone (BZ) gives the range of wave vectors **k** necessary to generate all distinct Bloch sums of atomic orbitals, without any duplication. In the free-electron model, it was also shown in Chapter 4 that the BZ boundaries give the **k** values where the electron waves are strongly perturbed by a periodic lattice potential. We shall explain here how to find the BZ in a general two- or three-dimensional lattice. It is easier to do this by using the LCAO definition, and then to show how the free-electron properties of the BZ follow.

In *one dimension*, the LCAO coefficient for atom n in a lattice of spacing a is:

$$c_n(k) = \exp(ikna) \tag{B.1}$$

If g is any multiple of $2\pi/a$:

$$g = 2p\pi/a\text{: } p = 0, \pm 1, \pm 2, \ldots \tag{B.2}$$

then

$$\begin{aligned} c_n(k+g) &= \exp(inka + i2pn\pi) \\ &= c_n(k). \end{aligned} \tag{B.3}$$

To generate all distinct $c_n(k)$, it is therefore sufficient to have k spanning a range of $2\pi/a$.

For the first BZ, we usually take the symmetrical range:

$$-\pi/a \leqslant k < \pi/a \tag{B.4}$$

as this shows the symmetry of the $E(k)$ functions more clearly.

The free electron waves are:

$$\psi_k(x) = \exp(ikx) = \cos(kx) + i\sin(kx) \tag{B.5}$$

and values of k at the first BZ boundary give sine and cosine waves which just match the lattice spacing, and so give an energy splitting as explained in Chapter 4. The same will happen whenever k is a multiple of π/a, and the equation:

$$k = \pm p\pi/a\text{: } p = 1, 2, 3, \ldots \tag{B.6}$$

defines further BZ boundaries, which are shown in Fig. B.1. The first BZ is taken as the fundamental range of k, used to plot all band structures. In the nearly-free electron theory, all k values are shifted into this range.

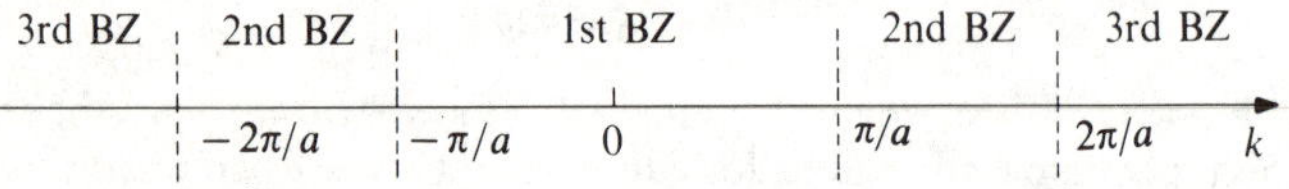

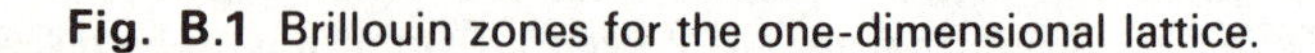

Fig. B.1 Brillouin zones for the one-dimensional lattice.

We now extend the above argument, which is a summary of that given in Chapter 4, to a two-dimensional lattice. A unit cell is defined by two vectors **a** and **b**, so that the position of a general lattice point, labelled (m, n), is:

$$\mathbf{r}_{m,n} = m\mathbf{a} + n\mathbf{b} \tag{B.7}$$

The free-electron functions in two dimensions are:

$$\psi_{\mathbf{k}}(\mathbf{r}) = \exp(i\mathbf{k}\cdot\mathbf{r}) \tag{B.8}$$

and by analogy, the Bloch sums in the LCAO theory are:

$$\psi_{\mathbf{k}}(\mathbf{r}) = \sum_{m,n} c_{m,n}(\mathbf{k})\,\chi_{m,n} \tag{B.9}$$

where:

$$c_{m,n}(\mathbf{k}) = \exp(i\mathbf{k}\cdot\mathbf{r}_{m,n}) \tag{B.10}$$

Now consider how to choose a vector **g**, which has the property:

$$c_{m,n}(\mathbf{k}+\mathbf{g}) = c_{m,n}(\mathbf{k}) \text{ for all } m, n. \tag{B.11}$$

We must have:

$$\exp(i\mathbf{g}\cdot\mathbf{r}_{m,n}) = 1 \tag{B.12}$$

or:

$$\mathbf{g}\cdot\mathbf{r}_{m,n} = 2p\pi \tag{B.13}$$

where p is an integer. We can satisfy this equation if **g** is an integral combination of two vectors **A** and **B** chosen so that:

$$\mathbf{A}\cdot\mathbf{a} = \mathbf{B}\cdot\mathbf{b} = 2\pi \tag{B.14}$$
$$\mathbf{A}\cdot\mathbf{b} = \mathbf{B}\cdot\mathbf{a} = 0.$$

Then if:

$$\mathbf{g} = h\mathbf{A} + k\mathbf{B} \tag{B.15}$$

we have:

$$\begin{aligned}\mathbf{g}\cdot\mathbf{r}_{m,n} &= (h\mathbf{A} + k\mathbf{B})\cdot(m\mathbf{a} + n\mathbf{b}) \\ &= 2\pi(hm + kn).\end{aligned} \tag{B.16}$$

A and **B** are called **reciprocal vectors**, and **g** is a point on a **reciprocal lattice**. (In crystallography, the factor 2π is normally omitted. However, the definition given here is the usual one in applications to band theory.)

Examples of reciprocal lattices corresponding to three different real lattices are shown in Fig. B.2. For a square lattice with side a, the reciprocal lattice is also square, with side $2\pi/a$. In the rectangular case, the reciprocal lattice has sides $2\pi/a$ and $2\pi/b$, corresponding to real lattice dimensions a and b respectively. When **a** and **b** are inclined, **A** will always be perpendicular to **b** and **B** to **a**. This is illustrated by the case of a

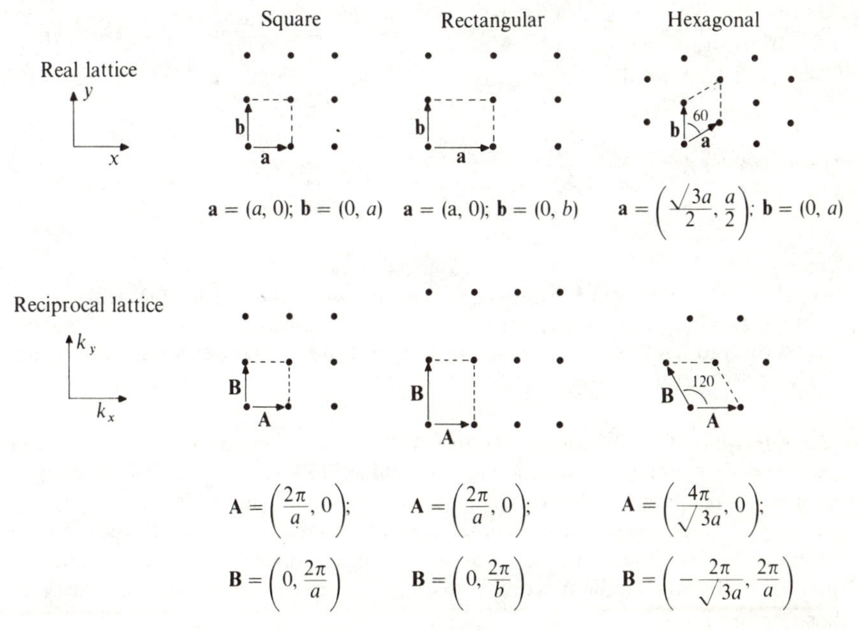

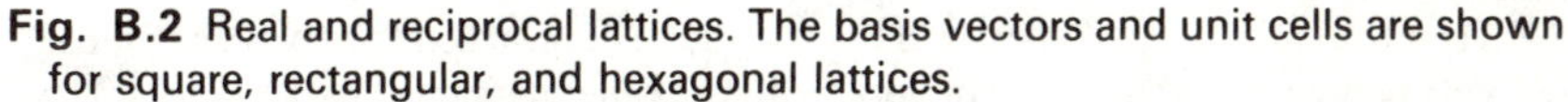

Fig. B.2 Real and reciprocal lattices. The basis vectors and unit cells are shown for square, rectangular, and hexagonal lattices.

hexagonal lattice, where the basis vectors drawn in Fig. B.2 are:

$$\mathbf{a} = (\sqrt{3}a/2,\ a/2) : \mathbf{b} = (0,\ a). \tag{B.17}$$

The corresponding reciprocal vectors are:

$$\mathbf{A} = (4\pi/\sqrt{3}a,\ 0) : \mathbf{B} = (-2\pi/\sqrt{3}a,\ 2\pi/a) \tag{B.18}$$

and also give a hexagonal lattice as shown.

The equation for **g** shows that if **k** is altered by any reciprocal lattice vector, the LCAO coefficients in the Bloch sum will be unchanged. To generate all distinct Bloch sums, therefore, we only need to take **k** over one unit cell of the reciprocal lattice. The normal unit cell, with its corner at the origin, does not reflect the symmetry of the lattice or of the $E(\mathbf{k})$ functions. We therefore take as the first BZ a *Wigner–Seitz* cell of the reciprocal lattice. This is the region enclosed by perpendicular bisectors of lines from the origin to all neighbouring reciprocal lattice points. Although in the square and rectangular lattices the BZ found in this way is also a square or a rectangle, in many lattices it has a different shape from the normally drawn primitive lattice cell. Thus in the hexagonal lattice, it is a hexagon, as shown in Fig. B.3. The construction of the Wigner–Seitz cell shows that every lattice point may be surrounded by an identical cell, in such a way as to fill the whole plane. Thus a Wigner–Seitz cell contains the same area as a normal lattice cell, although it may be arranged differently.

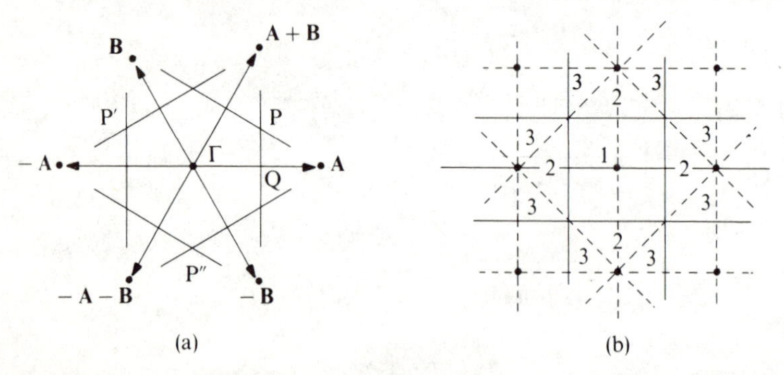

Fig. B.3 Construction of Brillouin zones in two dimensions. (a) Hexagonal lattice, showing the vectors to neighbouring reciprocal lattice points, labelled according to the basis vectors of Fig. B.2, and the hexagonal zone. (b) Square lattice, with further zone boundaries drawn.

The drawing of the hexagonal BZ in Fig. B.3 shows the points P and Q used in Chapter 4 to discuss the band structure of graphite. The **k** values given may be worked out from the reciprocal lattice vectors shown previously. It is interesting to note that the two points P′ and P″ in the diagram may be reached from P by reciprocal lattice vectors. It follows that Bloch sums with **k** vectors at these three points are identical. This gives a three-fold symmetry to the functions, which can be seen in the graphite orbitals at P, illustrated in Fig. 4.26 on p. 115.

As is shown for the square lattice in Fig. B.3, the BZ construction may be extended by drawing perpendicular bisectors of lines to more distant reciprocal lattice points. These define higher BZ, marked 2,3, . . . in the diagram. In the free-electron theory, values of **k** falling in these regions are shifted into the first BZ, so that all band structures can be plotted in this region.

We shall now show how the BZ construction outlined above relates to the behaviour of nearly-free electrons in the periodic potential of the lattice. A strong perturbation occurs whenever an integral number of half wavelengths just fits between adjacent rows of atoms. The relationship between **k** and wavelength shows that this will happen when the component of **k** perpendicular to the rows, k_n, is given by:

$$\pi/k_n = d \tag{B.19}$$

where d is the spacing between rows.

A row of atoms in the two-dimensional lattice can be described by the Miller indices (h, k), showing that it cuts the crystal axes at points $\mathbf{a}/h$ and $\mathbf{b}/k$. This is illustrated in Fig. B.4, where the row (1, 2) is shown. A vector in the same direction as the row (h, k) is:

$$\mathbf{v} = \mathbf{a}/h - \mathbf{b}/k. \tag{B.20}$$

Thus with the reciprocal lattice vector **g** defined above:

$$\mathbf{g}\cdot\mathbf{v} = (h\mathbf{A} + k\mathbf{B})\cdot(\mathbf{a}/h - \mathbf{b}/k) = 0 \tag{B.21}$$

This shows that a vector to the point (h, k) of the reciprocal lattice is perpendicular to the row (h, k) in the real lattice.

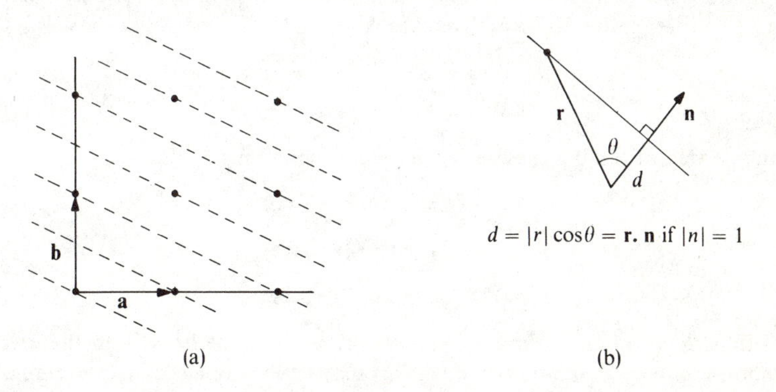

Fig. B.4 Rows in a two-dimensional lattice. (a) The rows drawn cross the axes at **a** and **b**/2, corresponding to the Miller indices (1, 2). (b) Illustration of the equation $\mathbf{r}\cdot\mathbf{n} = d$, where **r** is a vector to any point on a row from a point at distance d to it, and **n** is a unit vector perpendicular to the row.

The next step is to relate the length of **g** to the spacing d between adjacent (h, k) rows. Suppose that **n** is a vector of unit length, perpendicular to the row. If **r** is a vector from the origin to anywhere on the row, the construction illustrated in Fig. B.4(b) shows that:

$$\mathbf{n}\cdot\mathbf{r} = d \tag{B.22}$$

We choose **r** as some point, such as $\mathbf{a}/h$, which we know is on the row. Since **g** is perpendicular to the row, a suitable unit vector $\mathbf{n}$ is $\mathbf{g}/|g|$, where $|g|$ is the length of $\mathbf{g}$. Then:

$$(\mathbf{g}/|g|)\cdot(\mathbf{a}/h) = d$$

or:

$$d = 1/(h|g|)(h\mathbf{A}+k\mathbf{B})\cdot\mathbf{a}$$
$$= 2\pi/|g| \tag{B.23}$$

from the definitions of **A** and **B**.
Thus not only is the reciprocal lattice vector (h, k) perpendicular to the rows of the real lattice, but the length of this vector is inversely related to the spacing between rows.

Let us return to the BZ construction given above. Each boundary is the perpendicular bisector of some reciprocal lattice vector **g**. Thus if **k** is at a BZ boundary, its component in the direction of **g** is $|g|/2$. Since we have shown that **g** is perpendicular to a row of atoms with spacing d given above, we can see that the component of **k** perpendicular to the row is also:

$$k_{\mathrm{n}} = |g|/2 = \pi/d \tag{B.24}$$

This is just the condition for a strong interaction with the lattice.

The argument which has been given in some detail for two dimensions can be extended without difficulty to three. Now the crystal lattice is described by three basis vectors, **a**, **b**, and **c**. We define the reciprocal vectors **A**, **B**, and **C** by the conditions:

$$\mathbf{A}\cdot\mathbf{a} = \mathbf{B}\cdot\mathbf{b} = \mathbf{C}\cdot\mathbf{c} = 2\pi$$
$$\mathbf{B}\cdot\mathbf{a} = \mathbf{C}\cdot\mathbf{a} = \mathbf{A}\cdot\mathbf{b} = \ldots = 0 \tag{B.25}$$

A suitable way of finding these is by the vector equations:

$$\mathbf{A} = 2\pi(\mathbf{b}\times\mathbf{c})/(\mathbf{a}\cdot\mathbf{b}\times\mathbf{c})$$
$$\mathbf{B} = 2\pi(\mathbf{c}\times\mathbf{a})/(\mathbf{a}\cdot\mathbf{b}\times\mathbf{c})$$
$$\mathbf{C} = 2\pi(\mathbf{a}\times\mathbf{b})/(\mathbf{a}\cdot\mathbf{b}\times\mathbf{c}) \tag{B.26}$$

The BZ is now found by drawing planes which bisect the lines from the origin to neighbouring reciprocal lattice points. This may be illustrated for two examples (see Fig. B.5). In the simple cubic lattice, the reciprocal lattice is also cubic, with spacing $2\pi/a$. The first BZ is simply a cube surrounding the origin, giving the range of **k** values as:

$$-\pi/a \leqslant k_x, k_y, k_z < \pi/a \tag{B.27}$$

The f.c.c. lattice is slightly more complicated. The primitive basis vectors are shown in Fig. B.5(b). They are:

$$\mathbf{a} = (a/2, a/2, 0):\ \mathbf{b} = (0, a/2, a/2):\ \mathbf{c} = (a/2, 0, a/2) \tag{B.28}$$

The reciprocal lattice vectors can then be found to be:

$$\mathbf{A} = (2\pi/a)(1, 1, -1):\ \mathbf{B} = (2\pi/a)(-1, 1, 1):\ \mathbf{C} = (2\pi/a)(1, -1, 1) \tag{B.29}$$

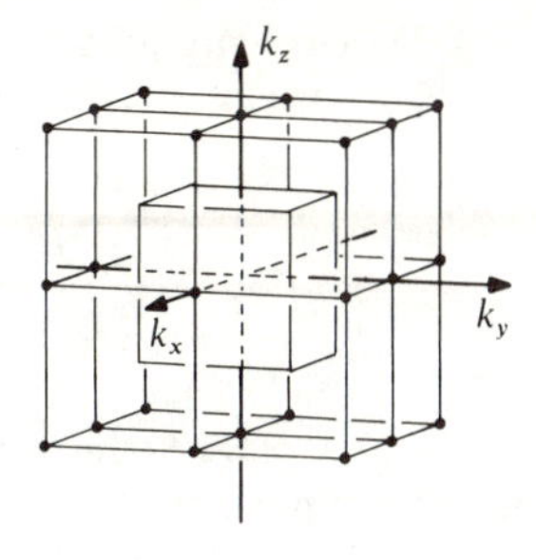

(a)

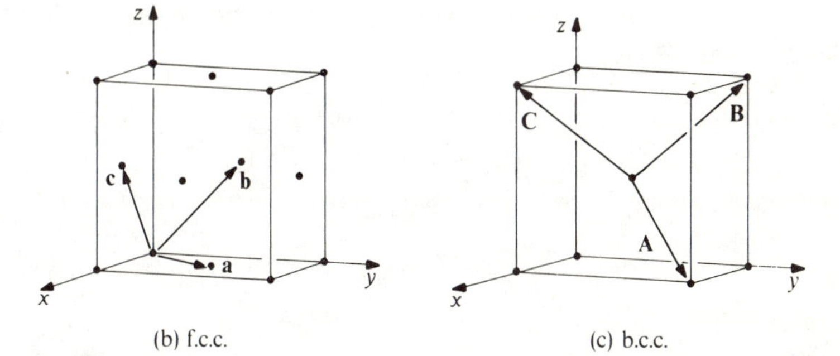

(b) f.c.c.

(c) b.c.c.

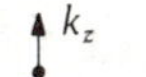

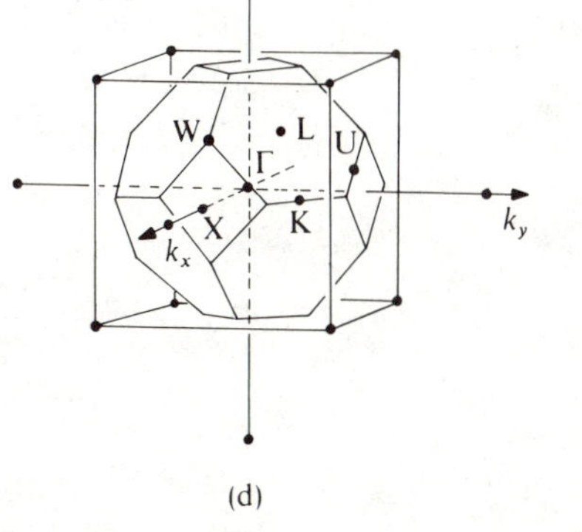

(d)

Fig. B.5 Brillouin zones in three-dimensional space. (a) BZ of a simple cubic lattice. (b) and (c): Primitive basis vectors for f.c.c. and b.c.c. lattices, respectively. (d) Brillouin zone of an f.c.c. lattice, constructed from the reciprocal b.c.c. lattice.

These are the basis vectors of a b.c.c. lattice, shown in Fig. B.5(c). Thus the first BZ is found by drawing planes which are perpendicular to the 14 ($= 6+8$) vectors to points near the origin in this lattice. This gives the truncated octahedron, drawn in

Fig. B.5(c), and used in Chapter 4. The conventional labels for different k values in the zone, which are used in plotting band-structure diagrams, are also shown.

Further reading

A detailed account of the construction of Brillouin zones for a variety of crystal structures, and of the application of group theory to the symmetries of electronic wave functions in solids, is given in:

H. Jones (1975). *The Theory of Brillouin zones and electronic states in crystals* (2nd edn). North-Holland.

Chemical formula index

Subject index